Lecture Notes in Physics

Lecture Notes in Physics

Edited by H. Araki, Kyoto, J. Ehlers, München, K. Hepp, Zürich
R. Kippenhahn, München, H. A. Weidenmüller, Heidelberg
and J. Zittartz, Köln

167

Gerhard Fieck

Symmetry of Polycentric Systems

The Polycentric Tensor Algebra for Molecules

Springer-Verlag
Berlin Heidelberg GmbH 1982

Author
Gerhard Fieck
Dannenberger Landstraße 10
2121 Dahlenburg

ISBN 978-3-540-11589-2   ISBN 978-3-540-39349-8 (eBook)
DOI 10.1007/978-3-540-39349-8

© by Springer-Verlag Berlin Heidelberg 1982

Originally published by Springer-Verlag Berlin Heidelberg New York in 1982

2153/3140-543210

To my wife

Contents

1. Introduction

Philosophers of science have traditionally been fascinated by the regular polyhedra. Pythagoras and Plato assigned them to the primary elements; Kepler speculated about them in "Harmonices Mundi"; and more recently W.Heisenberg, fully aware of this tradition, offers us a modern interpretation of Plato's opinion. Summing up, he says:

"Die letzte Wurzel der Erscheinungen ist also nicht die Materie, sondern das mathematische Gesetz, die Symmetrie, die mathematische Form." [1]

(The ultimate root of the phenomena is not the matter but the mathematical law, the symmetry, the mathematical form.)

Because the intention of this treatise hardly can be explained better, we add another quotation, in which we certainly may read molecule as well as elementary particle:

"Fragt man bei Plato, welches der Inhalt seiner Formen sei, aus welchem Stoff also seine regulären Körper schließlich gemacht seien, so erhält man die Antwort: aus Mathematik. Denn die Dreiecke, aus denen die regulären Körper gebildet werden sollen, sind ja nicht selbst Materie, da sie als zweidimensionale Gebilde keinen Raum erfüllen. Sie sind gedankliche Konstruktionen, die durch die Art ihrer Zusammenfügung räumliche Gebilde darstellen. In ähnlicher Weise sind in der heutigen Physik die Eigenwerte, die die Elementarteilchen darstellen, eben Eigenwerte einer Gleichung und in sofern rein mathematische Gebilde, denen keine Substanz zugrunde liegt. In gewisser Weise könnte man vielleicht noch die Energie als Substanz bezeichnen, aber auch die Energie und ihre Erhaltung ist eine mathematische Folge einer Invarianz-Eigenschaft der Gleichung, sie ist gewissermaßen in der Gleichung enthalten. Letzten Endes wird also der Materiebegriff in beiden Fällen auf Mathematik zurückgeführt. Der innerste Kern alles Stofflichen ist für uns wie für Plato eine "Form", nicht irgend ein materieller Inhalt. " [2]

(If one asks in Plato, what is the essence of his forms, i.e.which material his regular solids are made of, one gets the answer: of mathematics. For the triangles constituting the regular solids are no matter by themselves, because as two-dimensional entities they cover no space. They are constructions of thought, which constitute spatial formations by the mode of their composition. Similarly in modern physics the eigenvalues representing the elementary particles are just eigenvalues of an equation and thus purely mathematical entities based on no substrate. In a sense one might specify the energy as a substrate, but even the energy and its conservation is a mathematical consequence of an invariance of the equation, it is so to speak embodied in the equation. In the end in both cases, the concept of matter is reduced to mathemat-

ics. For us, as for Plato, the very essence of reality is a question of
"form", not of material substrate.)

The present treatise is devoted to the elaboration of such principles of form, with particular respect to molecular physics and quantum
chemistry. These principles appear on two levels. The classical symmetries of the polyhedra have been replaced in their importance by more
fundamental symmetries, which are shared by all elementary particles,
and consquently by all molecules. These symmetries, including the homo-
geneity and the isotropy of space-time and the permutational symmetry
according to Pauli's principle, specify the possible forms of Schrödinger equations and of state functions. These symmetries will be presup-
posed here, and will concern us only in the case of the multi-centre
integrals (section 13). For these integrals they prescribe general
principles of form not restricted to symmetric molecules.

The main subject of this treatise is the symmetry of the polyhedra
realized in "the architecture of molecules". Pauling's book of this
title [3] may be regarded as a modern illustration of Plato, and of
Kepler's statement, "geometria est archetypus pulchritudinis mundi".
The triangles mentioned by Plato as the constituents of the polyhedra
are essential to the present analysis, too.

The analysis of symmetry means application of group theory. Except
for the theory of transition-metal ions, this application in chemistry
has been more or less qualitative: labeling of states and normal vibrations, splitting and selection rules. The construction of symmetry-
adapted linear combinations (SALC),going beyond this point, has re-
mained unsystematic, because the linear combination coefficients - in
contrast to the Clebsch-Gordan coefficients - have not been considered
a basis of an algebra.

Against the "myth of qualitative group theory" [4], the Wigner-
Eckart theorem in its several varieties yields qualitative results and
makes the quantitative analysis of symmetry a theory of reduced matrix
elements or, more generally speaking, a theory of invariants. In atomic spectroscopy, the reduced matrix elements are related to the radial
integrals of the atomic orbitals, the Slater integrals for instance,
and therefore have a lucid meaning. The same holds for the ligand field
theory, which is focussed on the central transition-metal ion. But what
is the concrete meaning of the reduced matrix elements in polycentric
systems?

As an answer to this question, we shall design a quantitative analysis of symmetry for molecules, which can not be treated as quasi-mono-
atomic systems. Consequently, this must be a theory of molecular invariants. Since the symmetric coordination polyhedra are a new compo-

nent in comparison to the traditional atomic spectroscopy, to the nuclear shell and to the ligand field theory, this will exercise an influence on the character of molecular invariants as well as on the algebra of coefficients.

With regard to the invariants there are now two different types. At one hand there are still the reduced matrix elements of the Wigner-Eckart theorem. Because these arise from the s.-a. and thus delocalized molecular functions, they gain their significance only indirectly by as second type of invariants localized at the edges, triangles etc. of the polyhedral framework. These invariants involving several atomic centres quite naturally refer to the neighbourhoods within the co-ordination polyhedra (sections 4, 8, 11) and thus give way to the ideas of coordination chemistry and the theory of chemical binding.

With respect to the theorems mediating the connections of both types of invariants, we need a polyhedral supplement of the tensor algebra. The familiar classes of coefficients in the Wigner-Racah algebra (3jm, 6j, 9j symbols, isoscalars, coefficients of reduction and fractional parentage) are supplemented by classes with reference to the edges, triangles and deformed tetrahedra subtended by the atoms. The most significant, new coefficients represent triangular relations within the polyhedra. In analogy to the familiar isoscalars, they are called polyhedral isoscalars. Principally the new definitons and the somewhat delicate design of the graphic symbols shall be kept as close as possible to the existing Wigner-Racah algebra.

The question for molecular invariants can be formalized quite generally in a basic concept. An arbitrary statement of the ligand field theory may serve as an example, let us say the energy of the state $^1A_{1g}(t^2_{2g})$ in an octahedral field:

$$E(^1A_{1g}(t^2_{2g})) = E_0 - 8Dq + A + 10B + 5C \tag{1.1}$$

On the left we have a physical quantity $Ph(y)$ depending on a set of quantum numbers y. On the right there are several radial integrals $Int(z)$ likewise depending on quantum numbers z. In (1.1) these are indicated by distinct letters. Seldom is it stressed that the factors -8, 1, 10, and 5 are not more or less occasional numbers but values of functions of y and z and therefore can be enunciated by formulas containing the sets y and z. Factors of this type will be called geometrical factors through out, $GEO(y,z)$ in this case. (1.1) now presents itself as a special case of basic relations:

$$Ph(y) = \sum_z GEO(y,z) \cdot Int(z) \tag{1.2}$$

Another example is the square of a vibrational frequency of a molecule expressed by the force constants with respect to neighbouring atoms. This relation is determined by geometrical factors, too.

Often - especially for symmetrical structures - the geometrical factors are the only accessible, precise data and therefore the most interesting part of the theory. In semi-empirical theories (like Hückel-MO), the integrals $Int(z)$ are treated as parameters. For the fitting of these parameters, the inversion

$$Int(z) = \sum_y GEO(y,z) \cdot Ph(y) \tag{1.3}$$

is useful.

The task now is to design a systematic theory of the quantities $Int(z)$ at one hand and the geometrical factors at the other, as has been elaborated for atoms and nuclei.

The physical quantities naturally do not depend on the choice of coordinate axes and the numbering of atoms; they are invariants. The generation of a minimum number of invariants $Int(z)$ demands the consideration of symmetry. Since (1.2) is a relation between invariants of different type, there can occur nested relations of this kind. An example is a many-particle matrix element, which, in a first step, is expressed by the one-particle MO-integrals (respectively their invariants). In a second step, the latter are expressed by the multi-centre integrals of the atomic functions (respectively their invariants).

The geometrical factors, being invariants too, have to be reduced to the coefficients of the Wigner-Racah algebra of the relevant point group and its polycentric supplement. The generalization of the Wigner-Racah algebra to point groups has been given by Griffith [5] with restriction to simply reducible groups containing ambivalent classes. Even more recent books in this field cling to this restriction [6, 7, 8]. An exception is the introduction to the energy diagrams published by König and Kremer [9] and preceeding papers by the same authors. Since our results shall equally and immediately apply to systems like $Co(NO_2)_6^{3-}$ (group T_h), Ta_6-clusters with spin-orbit coupling (group O_h^*), $Pt(CN)_4^{2-}$ in an magnetic field (group C_{4h}), and $B_{12}H_{12}^{2-}$ (icosahedral group), we include the non-simply-reducible point groups with non-ambivalent classes from the beginning.

Because of the heterogeneous notations in the literature, we recapitulate the Wigner-Racah algebra of non-simply-reducible point groups with non-ambivalent classes in section 2. The price for including these groups is the appearence of several multiplicity indices, which may obscure the esthetic clarity of the theorems. Therefore it is advisable to omit all multiplicity indices and their related sums from time to time. Mostly these indices take only one value and are included only as a precaution for the few cases of multiplicity 2 or 3. After this recapitulation, the principal concepts refering to point group symmetry only are built up in the sections 3 to 11. In doing so, the preliminary studies [10, 11] are resumed. The rest of this treatise is new matter. The sections from 12 onwards contain more detailed results, which depend more on special choices concerning multiplicities and orbital systems.

Since many-electron matrix elements are reduced to one- or two-electron matrix elements by geometric relations, we are mainly concerned with the latter. Because of the delocalization of the MOs, the manyelectron systems in the MO scheme show no polycentric peculiarities compared to ligand field theory. So they are dealt with only in short in section 19. On the contrary, the VB scheme, though less usual, leads to some general ideas in our context (section 20). The main subject of the further development presumably is the theory of the polycentric coefficients of fractional parentage (the generalization of the familiar coefficients). For this purpose, one needs the unitary group or quasi-spin approach adopted by Racah for the many-particle theory.

Another direction of improvement is sketched in section 21. It is the application of the present symmetry-analysis on crystals, i.e. the introduction of space groups into our considerations. This application offers itself in particular for calculations using atomic or Wannier functions (for instance tight-binding or OPW methods) and for normal vibrations of crystals. In the case of symmorphic space-groups this introduction causes no difficulties, but for the non-symmorphic groups it is hampered by the deficient elaboration of the Wigner-Racah algebra of these groups.

Another aspect, the application to the molecular normal vibrations, has been demonstrated in the paper [12] . Since no principal new concepts could be expected to appear, it has not been taken up again here. The same applies to the more complicated adiabatic and relativistic effects.

In the foreword of his book [13], Chesnut has termed group theory as organized common sense. Organizing in this context also means the tabulation or the programming of the group-theoretical coefficients, just as they now are available for the rotation group [14] and by a recent book [75] for the point groups. Not until the newly defined coefficients, especially the polyhedral isoscalars, are available numerically, the symmetry-analysis lined out here will take full effect.

2. Summary of the Wigner-Racah algebra of non-simply-reducible point groups with non-ambivalent classes

In the following, we assume as known the results of the monocentric group theory of atoms and molecules presented, let us say, in the books of Edmonds [15] and Griffith [5]. Some parts of the group theory concerning the two-particle interactions and coefficients of fractional parentage (CFP) and missing in Griffith's book have been described in the papers [16, 17] and this concise summary is in part identical with the appendix of [17], where more references are given. A detailed study of the NSR groups has been made by Butler [18], but it is more general than necessary for point groups. We further refer to the lectures given by Butler and Piepho at the NATO Advanced Study Institute on recent advances in group theory [19, 20].

2.1. Representations

We consider a molecule, complex, or cluster with a symmetry group G, optionally a point group, a double point group, or the direct product of a point group and the spin group $SU(2)$. The classes of G are denoted by C and its elements by g, g' etc. The latter are identified with the symmetry operations in the configuration space acting on vectors, triangles and other objects in space, for instance:

$$\vec{r}' = g^{-1}\vec{r} \tag{2.1}$$

The irreducible representations of G are denoted by a, b, c etc. (or $a(G)$, $b(G)$, ... if a distinction is necessary) with the unitary representation matrices $D^a(g)$, and the matrix elements $D^a_{mn}(g)$:

$$D^a(g)D^a(g) = D^a(gg') \text{ or } \sum_n D^a_{mn}(g)D^a_{np}(g') = D^a_{mp}(gg') \tag{2.2}$$

$$D^a(g)^+ = D^a(g)^{-1} = D^a(g^{-1}) \text{ or}$$
$$\left[D^a(g)^+\right]_{mn} = \left[D^a_{nm}(g)\right]^* = \left[D^a(g)^{-1}\right]_{mn} = D^a_{mn}(g^{-1}) \tag{2.3}$$

These obey the orthogonality relation:

$$\sum_{g \in G} D^a_{mn}(g)D^b_{pq}(g)^* = \frac{ordG}{dima} \cdot \delta_{ab}\delta_{mp}\delta_{nq} \tag{2.4}$$

The representation contragredient to a is denoted by a^+ and defined by

$$D^{(a^+)}_{mn}(g) = \left[D^a_{mn}(g)\right]^* = D^a_{nm}(g^{-1}). \tag{2.5}$$

The reducible finite dimensional representations of G are denoted by σ^A (or $\sigma^A(G)$) with a discriminating index A and the representation by unitary operators, acting on a Hilbert space, by $U(g)$. The operators $U(g)$ are defined in the space representation by

$$\langle g^{-1}\vec{r}|\phi\rangle = \langle \vec{r}|U(g)|\phi\rangle. \tag{2.6}$$

2.2. Characters

The characters are defined as usual for irreducible and reducible representations:

$$\chi^a(g) = \sum_m D^a_{mm}(g), \qquad \sigma^X(g) = \sum_i \sigma^X_{ii}(g) \tag{2.7}$$

From (2.4) follows the orthogonality relation

$$\text{ord}G^{-1}\sum_{C \in G}\text{ord}C \cdot \chi^a(C)\chi^{b*}(C) = \delta_{ab}, \tag{2.8}$$

where C are the classes of G and ordC their order, i.e. the number of elements in C. A further orthogonality relation is:

$$\text{ord}G^{-1}\sum_a \chi^a(C)\chi^{a*}(C') = \text{ord}C^{-1}\delta_{CC'} \tag{2.9}$$

The criterion

$$n(X,a) = \text{ord}G^{-1}\sum_C \text{ord}C \cdot \chi^a(C)^* \cdot \sigma^X(C) \tag{2.10}$$

gives the multiplicity in the decomposition, i.e. the direct sum

$$\sigma^X = \sum_a n(X,a) \cdot a \tag{2.11}$$

The relation

$$\text{ord}G^{-1}\sum_C \text{ord}C \cdot \chi^a(C)\chi^b(C)\chi^c(C) = n(abc) \tag{2.12}$$

likewise gives the multiplicity in the decompositions of the direct products:

$$a \times b = \sum_c n(abc)c^+, \quad a \times c = \sum_b n(abc)b^+, \text{ and } b \times c = \sum_a n(abc)a^+ \tag{2.13}$$

But n(abc) also determines the multiplicity of the identical representation in the triple product $a \times b \times c$. If $n(abc) \geqslant 1$ (abc) is called a triad. In order to discriminate triads from other triples of representations, one defines the symbol:

$$\delta(abc) = \begin{cases} 1 & \text{if } n(abc) \geqslant 1 \\ 0 & \text{if } n(abc) = 0 \end{cases} \tag{2.14}$$

In the case of $n(abc) \geqslant 2$, we need a multiplicity index in the decompositions (2.13), τ let us say, and shall call (abcτ) a triad too. Because of (2.8) we have for all groups and all irreducible representations a: $n(aa^+1) = 1$.

Concerning the equivalence of a representation a to a^+ or to a real representation a_r, there are three cases:

$$\text{a) } a \sim a^+ \sim a_r, \qquad \text{b) } a \not\sim a^+, \qquad \text{c) } a \sim a^+ \not\sim a_r$$

These cases are distinguished by the criterion:

$$\sum_{g \in G}\chi^a(g^2) = \begin{cases} 1 & \text{in case a} \\ 0 & \text{in case b} \\ -1 & \text{in case c} \end{cases} \tag{2.15}$$

In the cases a) and c), the characters of the representations are real.

In the tables of Koster e.a. [21], the cases are specified for each representation.

Finally, all point groups are simple phase groups, which satisfy:

$$\sum_{g \in G} \chi^a(g^3) = \sum_{g \in G} \chi^a(g)^3 \tag{2.16}$$

This means that the identical representation may be contained only in
the totally symmetric or the totally anti-symmetric part of the direct
triple product $a \times a \times a$. The significance of this property will show up
in section 2.4.

2.3. Bases

The bases carrying the representation a are denoted by am (and a discriminating multiplicity index α if necessary). If the basis is realized
by Hilbert space vectors, we use Diracs bra-ket notation through out:

$$U(g)|...am\rangle = \sum_n D^a_{nm}(g) \cdot |...an\rangle \tag{2.17}$$

In the case of finite-dimensional vector spaces, except the tree-dimen-
sional configuration space, we use $|...am)$ in the same sense.

The basis contragredient to $|am\rangle$ is denoted by $|a^+m\rangle$ having the same
transformation properties as the bras:

$$\langle \bar{r}|am\rangle = \langle am|\bar{r}\rangle^* = \langle a^+m|\bar{r}\rangle \quad \text{or} \quad (F|am) = (am|F)^* = (a^+m|F)$$

Since a and a^+ may be inequivalent, we can not relate am and a^+m in
general. Moreover, such a relation is of marginal interest, because the
final invariant expressions do not contain special bases anymore. But
the basis $a^{++}m$ has the same transformational property as am and therefore is proportional to the latter. A detailed consideration leads to

$$a^{++}m = \{a\} \cdot am \tag{2.18}$$

with a phase factor $\{a\} = 1$ for vector and $\{a\} = -1$ for spinor representations [22]. The use of braces for this phase factor is justified by
the relations (2.31) and (2.39). The representations of case a) are all
vector representations, those of case c) all spinor representations.
The complex representations of case b) can belong to either type.

We have to take into account (2.18), if we relabel representations:

$$am \rightarrow b^+m \quad \text{implies} \quad a^+m \rightarrow b^{++}m = \{b\} bm$$

and $\quad a^+m \rightarrow bm \quad \text{implies} \quad am \rightarrow b^{+++}m = \{b\} b^+m$

For all point groups, the following quasi-ambivalence condition [19]
holds (cf.also eq.(12)of [20]):

$$\{a\}\{b\}\{c\} = 1 \text{ if } (abc) \text{ is a triad, i.e. } n(abc) \geq 1 \tag{2.19}$$

Since there is a strict correspondence of co- and contragredient
bases in all sums, we can adopt Einsteins convention: If the same in-

dex occurs twice, as a co- and a contragredient index, the the sum is
to be taken over the range of this index:

$$\sum(\cdots {\overset{+}{\underset{m}{a}}} \cdots)(\cdots {\underset{m}{a}} \cdots) = \sum_m (\cdots {\overset{+}{\underset{m}{a}}} \cdots)(\cdots {\underset{m}{a}} \cdots) \qquad (2.20)$$

2.4. 3jm symbols

The matrix

$$M^{abc}_{lmn,pqr} = \text{ordG}^{-1} \sum_{g \in G} D^a_{lp}(g) D^b_{mq}(g) D^c_{nr}(g) \qquad (2.21)$$

projects out the identical representation from the direct triple product $a \times b \times c$. Because of the relation $M^2 = M$, the eigenvalues of M are
zero and one. The number of the eigenvectors belonging to 1 is given
by $n(abc)$ according to (2.12). The components of these eigenvectors
(numbered by the multiplicity index σ) are the 3jm symbols of the group
G:

$$\text{ordG}^{-1} \sum_g D^a_{lp}(g) D^b_{mq}(g) D^c_{nr}(g) \cdot \binom{abc}{lmn}^{\sigma} = \binom{abc}{pqr}^{\sigma} \qquad (2.22)$$

For convenience the eigenvectors are chosen orthogonal:

$$\sum \binom{abc}{lmn}^{\sigma *} \binom{abc}{lmn}^{\tau} = \delta(\sigma, \tau) \qquad (2.23)$$

Since a matrix can be represented in terms of its eigenvectors and
-values, another definition of the 3jm symbols is

$$\text{ordG}^{-1} \sum_g D^a_{lp}(g) D^b_{mq}(g) D^c_{nr}(g) = \sum_\sigma \binom{abc}{lmn}^{\sigma *} \binom{abc}{pqr}^{\sigma}. \qquad (2.24)$$

The basic transformational property of the 3jm symbols is

$$\sum D^a_{lp}(g) D^b_{mq}(g) D^c_{nr}(g) \binom{abc}{lmn}^{\sigma} = \binom{abc}{pqr}^{\sigma} , \qquad (2.25)$$

from which other forms can be derived by (2.2). Because of the choice
(2.23), the orthogonality relations read:

$$\sum \binom{abc}{lmn}^{\sigma *} \binom{abd}{lmp}^{\tau} = \delta(c,d)\delta(\sigma,\tau)\delta(n,p)/\text{dimc} \qquad (2.26)$$

$$\sum_{\sigma cn} \text{dimc} \binom{abc}{lmn}^{\sigma *} \binom{abc}{pqn}^{\sigma} = \delta(l,p)\delta(n,q) \qquad (2.27)$$

In [18] Butler shows, that (2.19) allows the choice of the following
simple conjugation property:

$$\binom{abc}{lmn}^{\sigma *} = \binom{a^+ b^+ c^+}{l \ m \ n}^{\sigma} \qquad (2.28)$$

Because of the simple phase condition (2.16), all 3jm symbols can be
arranged according to the following symmetry rules. They are symmetric
with respect to cyclic permutations

$$\binom{abc}{lmn}^{\sigma} = \binom{bca}{mnl}^{\sigma} = \binom{cab}{nlm}^{\sigma} \qquad (2.29)$$

and need a phase factor $\{abc\sigma\} = \pm 1$ for odd permutations:

$$\begin{pmatrix}abc\\lmn\end{pmatrix}^{\sigma} = \{abc\sigma\}\begin{pmatrix}bac\\mln\end{pmatrix}^{\sigma} = \{abc\sigma\}\begin{pmatrix}acb\\lnm\end{pmatrix}^{\sigma} = \{abc\sigma\}\begin{pmatrix}cba\\nml\end{pmatrix}^{\sigma} \tag{2.30}$$

In the case of three different representations a, b, c, the choice of these phase factors is free. The factors $\{aaa\sigma\}$ are fixed, the factos $\{aac\sigma\}$ fixed in part. For SR groups they can be split up into phases associated to the individual representations (cf.eq.(7) of [20]):

$$\{abc\} = (-1)^{a} \cdot (-1)^{b} \cdot (-1)^{c}$$

The phase factors of (2.18) are special cases:

$$\{a\} = \{aa^{+}1\} \tag{2.31}$$

From (2.28), (2.18) and (2.19) we get the derived conjugation property:

$$\begin{pmatrix}abc^{+}\\lmn\end{pmatrix}^{\sigma *} = \{c\}\begin{pmatrix}a^{+}b^{+}c\\l\ m\ n\end{pmatrix}^{\sigma} = \{a\}\{b\}\begin{pmatrix}a^{+}b^{+}c\\l\ m\ n\end{pmatrix}^{\sigma} \tag{2.32}$$

If the triad contains the identical representation 1, we have the special case:

$$\begin{pmatrix}a1c\\k11\end{pmatrix}^{\varepsilon} = \delta(\varepsilon,1)\delta(a,c^{+})\delta(k,1)/\sqrt{dima} \tag{2.33}$$

In terms of the 3jm symbols, the coupling of kets is given by

$$|(ab)\sigma cp\rangle = \{c\}\sqrt{dimc}\sum\begin{pmatrix}a^{+}b^{+}c\\m\ n\ p\end{pmatrix}^{\sigma} \cdot |am\rangle \cdot |bn\rangle \tag{2.34}$$

The Clebsch-Gordan coefficients in Butler's "sensible" choice [18] therefore are :

$$\langle am,bn|(ab)\sigma cp\rangle_{s} = \{c\}\sqrt{dimc}\begin{pmatrix}a^{+}b^{+}c\\m\ n\ p\end{pmatrix}^{\sigma}$$

But there may be introduced an arbitrary phase factor K(abcσ):

$$\langle am,bn|(ab)\sigma cp\rangle = K(abc\sigma) \cdot \langle am,bn|(ab)\sigma cp\rangle_{s}.$$

We therefore shall use only the explicit formula (2.34).

2.5. 6j symbols

The 6j symbols are the invariants defined by:

$$\begin{Bmatrix}abc\\def\end{Bmatrix}_{\alpha\beta\gamma\delta} = \sum\begin{pmatrix}ae^{+}f\\il\ m\end{pmatrix}^{\alpha} \cdot \begin{pmatrix}dbf^{+}\\njm\end{pmatrix}^{\beta} \cdot \begin{pmatrix}d^{+}ec\\n\ lk\end{pmatrix}^{\gamma} \cdot \begin{pmatrix}a^{+}b^{+}c^{+}\\i\ j\ k\end{pmatrix}^{\delta} \tag{2.35}$$

Their symmetries follow from (2.29/30):

$$\begin{Bmatrix}abc\\def\end{Bmatrix}_{\alpha\beta\gamma\delta} = \begin{Bmatrix}cab\\fde\end{Bmatrix}_{\gamma\alpha\beta} = \begin{Bmatrix}bca\\efd\end{Bmatrix}_{\beta\gamma\alpha\delta}, \tag{2.36}$$

$$\begin{Bmatrix}abc\\def\end{Bmatrix}_{\alpha\beta\gamma\delta} = \{d\}\{e\}\{f\}\{afe^{+}\alpha\}\{df^{+}b\beta\}\{d^{+}ce\gamma\}\{abc\delta\}\begin{Bmatrix}a\ c\ b\\d^{+}f^{+}e\end{Bmatrix}_{\alpha\gamma\beta\delta} \tag{2.37}$$

and its combinations with (2.36).

$$\begin{Bmatrix}abc\\def\end{Bmatrix}_{\alpha\beta\gamma\delta} = \begin{Bmatrix}a^{+}e\ f^{+}\\d^{+}b\ c^{+}\end{Bmatrix}_{\delta\gamma\beta\alpha} = \begin{Bmatrix}d^{+}b^{+}f\\a^{+}e^{+}c\end{Bmatrix}_{\gamma\delta\alpha\beta} = \begin{Bmatrix}d\ e^{+}c^{+}\\a\ b^{+}f^{+}\end{Bmatrix}_{\beta\alpha\delta\gamma} \tag{2.38}$$

Because of (2.33) the 6j symbol containing the identical representation reduces to:

$$\begin{Bmatrix}abc\\de1\end{Bmatrix}_{11\gamma\delta} = \delta(\gamma,\delta)\delta(a,e)\delta(b^{+},d)\{abc\delta\}/\sqrt{dima \cdot dimb} \tag{2.39}$$

This formula justifies the choice of curly brackets for the phase factors in (2.30) and (2.18).

The essential relation of the 6j symbols is:

$$\sum \left(\begin{smallmatrix}ae^+f\\il\ m\end{smallmatrix}\right)^\alpha \cdot \left(\begin{smallmatrix}dbf^+\\njm\end{smallmatrix}\right)^\beta \cdot \left(\begin{smallmatrix}d^+ec\\n\ lk\end{smallmatrix}\right)^\gamma \cdot \left(\begin{smallmatrix}a^+b^+g^+\\i\ j\ p\end{smallmatrix}\right)^\delta = \delta(c,g)\delta(k,p)\mathrm{dimc}^{-1}\cdot\left\{\begin{smallmatrix}abc\\def\end{smallmatrix}\right\}_{\alpha\beta\gamma\delta} \quad (2.40)$$

This implies the definition (2.35) as a special case.

From (2.40) further relations can be derived by the help of (2.26/27):

$$\sum \left(\begin{smallmatrix}ae^+f\\il\ m\end{smallmatrix}\right)^\alpha \cdot \left(\begin{smallmatrix}dbf^+\\njm\end{smallmatrix}\right)^\beta \cdot \left(\begin{smallmatrix}d^+ec\\n\ lk\end{smallmatrix}\right)^\gamma = \sum_\delta \left\{\begin{smallmatrix}abc\\def\end{smallmatrix}\right\}_{\alpha\beta\gamma\delta}\left(\begin{smallmatrix}abc\\ijk\end{smallmatrix}\right)^\delta \quad (2.41)$$

and the recoupling equation:

$$\sum \left(\begin{smallmatrix}ae^+f\\il\ m\end{smallmatrix}\right)^\alpha \cdot \left(\begin{smallmatrix}dbf^+\\njm\end{smallmatrix}\right)^\beta = \sum_{\gamma\delta c}\mathrm{dimc}\left\{\begin{smallmatrix}abc\\def\end{smallmatrix}\right\}_{\alpha\beta\gamma\delta}\left(\begin{smallmatrix}abc\\ijk\end{smallmatrix}\right)^\delta \cdot \left(\begin{smallmatrix}d^+ec\\n\ lk\end{smallmatrix}\right)^{\gamma*} \quad (2.42)$$

For the elimination of the complex conjugation on the right, one has to take into account (2.32).

The relations containing several 6j symbols are as follows:

Orthogonality:

$$\sum_{\alpha\beta f}\mathrm{dimf}\left\{\begin{smallmatrix}abc\\def\end{smallmatrix}\right\}_{\alpha\beta\gamma\delta}^* \cdot \left\{\begin{smallmatrix}abg\\def\end{smallmatrix}\right\}_{\alpha\beta\sigma\tau} = \delta(c,g)\delta(\gamma,\sigma)\delta(\delta,\tau)/\mathrm{dimc} \quad (2.43)$$

Racah s back-coupling rule:

$$\sum_{\varepsilon\eta g}\mathrm{dimg}\{e\}\{dbf^+\beta\}\{abc\delta\}\{adg^+\eta\}\left\{\begin{smallmatrix}bac\\deg\end{smallmatrix}\right\}_{\varepsilon\eta\gamma\delta}\cdot\left\{\begin{smallmatrix}adg^+\\bef\end{smallmatrix}\right\}_{\alpha\beta\varepsilon\eta} = \left\{\begin{smallmatrix}abc\\def\end{smallmatrix}\right\}_{\alpha\beta\gamma\delta} \quad (2.44)$$

Biedenharn-Elliott sum rule:

$$\sum_\gamma \{a_1a_2a_3\gamma\}\left\{\begin{smallmatrix}a_1a_2a_3\\b_1b_2b_3\end{smallmatrix}\right\}_{\alpha_1\alpha_2\alpha_3\gamma}\left\{\begin{smallmatrix}a_3a_2a_1\\c_3c_2c_1\end{smallmatrix}\right\}_{\beta_3\beta_2\beta_1\gamma} = \sum_{a\alpha\beta\gamma}\{a_3^2\}\mathrm{dima}$$

$$\cdot\{ab_1c_1\alpha\}\{ab_2c_2\beta\}\{ab_3c_3\gamma\}\left\{\begin{smallmatrix}c_2b_2a\\b_1c_1a_3^+\end{smallmatrix}\right\}_{\beta_3\alpha_3\alpha\beta}\left\{\begin{smallmatrix}c_1b_1a\\b_3b_3a_2^+\end{smallmatrix}\right\}_{\beta_2\alpha_2\gamma\alpha}\left\{\begin{smallmatrix}c_3b_3a\\b_2c_2a_1^+\end{smallmatrix}\right\}_{\beta_1\alpha_1\beta\gamma} \quad (2.45)$$

Special cases of these relations are the sum rules over one 6j symbol:

$$\sum_{\gamma c}\mathrm{dimc}\left\{\begin{smallmatrix}abc\\abf\end{smallmatrix}\right\}_{\alpha\beta\gamma\gamma} = \delta(\alpha,\beta)\delta(ab^+f\alpha) \quad (2.46)$$

$$\sum_{\gamma c}\{abc\gamma\}\mathrm{dimc}\left\{\begin{smallmatrix}abc\\baf\end{smallmatrix}\right\}_{\alpha\beta\gamma\gamma} = \delta(f,1)\delta(\alpha,1)\delta(\beta,1)\sqrt{\mathrm{dima}\cdot\mathrm{dimb}} \quad (2.47)$$

2.6. The 9j symbols

We now collect the properties of the 9j symbols. They are defined by

$$\left\{\begin{smallmatrix}a\ b\ c\\d\ e\ f\\g\ h\ i\end{smallmatrix}\right\}^\alpha_{\beta\gamma} = \sum \left(\begin{smallmatrix}a\ b\ c\\m_a m_b m_c\end{smallmatrix}\right)^{\alpha*}\left(\begin{smallmatrix}d\ e\ f\\m_d m_e m_f\end{smallmatrix}\right)^{\beta*}\left(\begin{smallmatrix}g\ h\ i\\m_g m_h m_i\end{smallmatrix}\right)^{\gamma*}\left(\begin{smallmatrix}a\ d\ g\\m_a m_d m_g\end{smallmatrix}\right)^\delta\left(\begin{smallmatrix}b\ e\ h\\m_b m_e m_h\end{smallmatrix}\right)^\varepsilon\left(\begin{smallmatrix}c\ f\ i\\m_c m_f m_i\end{smallmatrix}\right)^\eta \quad (2.48)$$

and are invariant to even permutations of rows and columns. The multiplicity indices follow their triads in an obvious way:

$$\begin{Bmatrix} a & b & c \\ d & e & f \\ g & h & i \end{Bmatrix}\begin{matrix}\alpha \\ \beta \\ \gamma\end{matrix}\begin{matrix} \\ \\ \\ \delta\ \epsilon\ \eta\end{matrix} = \begin{Bmatrix} b & c & a \\ e & f & d \\ h & i & g \end{Bmatrix}\begin{matrix}\alpha \\ \beta \\ \gamma\end{matrix}\begin{matrix} \\ \\ \\ \epsilon\ \eta\ \delta\end{matrix} = \begin{Bmatrix} d & e & f \\ g & h & i \\ a & b & c \end{Bmatrix}\begin{matrix}\beta \\ \gamma \\ \alpha\end{matrix}\begin{matrix} \\ \\ \\ \delta\ \epsilon\ \eta\end{matrix} \qquad \text{etc.} \tag{2.49}$$

In the case of odd permutations of rows or columns, again, the phase factors appear:

$$\begin{Bmatrix} a & b & c \\ d & e & f \\ g & h & i \end{Bmatrix}\begin{matrix}\alpha \\ \beta \\ \gamma\end{matrix}\begin{matrix} \\ \\ \\ \delta\ \epsilon\ \eta\end{matrix}=\{abc\}\{def\beta\}\{ghi\gamma\}\begin{Bmatrix} a & c & b \\ d & f & e \\ g & i & h \end{Bmatrix}\begin{matrix}\alpha \\ \beta \\ \gamma\end{matrix}\begin{matrix} \\ \\ \\ \delta\ \eta\ \epsilon\end{matrix}=\{adg\delta\}\{beh\epsilon\}\{cfi\eta\}\begin{Bmatrix} a & b & c \\ g & h & i \\ d & e & f \end{Bmatrix}\begin{matrix}\alpha \\ \gamma \\ \beta\end{matrix}\begin{matrix} \\ \\ \\ \delta\ \epsilon\ \eta\end{matrix} \tag{2.50}$$

The conjugation is related to the reflection about the main diagonal:

$$\begin{Bmatrix} a^+ & b^+ & c^+ \\ d^+ & e^+ & f^+ \\ g^+ & h^+ & i^+ \end{Bmatrix}\begin{matrix}\alpha \\ \beta \\ \gamma\end{matrix}\begin{matrix} \\ \\ \\ \delta\ \epsilon\ \eta\end{matrix} = \begin{Bmatrix} a & b & c \\ d & e & f \\ g & h & i \end{Bmatrix}\begin{matrix}\alpha^* \\ \beta \\ \gamma\end{matrix}\begin{matrix} \\ \\ \\ \delta\ \epsilon\ \eta\end{matrix} = \begin{Bmatrix} a & d & g \\ b & e & h \\ c & f & i \end{Bmatrix}\begin{matrix}\delta \\ \epsilon \\ \eta\end{matrix}\begin{matrix} \\ \\ \\ \alpha\ \beta\ \gamma\end{matrix} \tag{2.51}$$

As all nj symbols, the 9j symbols containing the identical representation reduce to zero or a simpler symbol:

$$\begin{Bmatrix} a & b & 1 \\ d & e & f \\ g & h & i \end{Bmatrix}\begin{matrix}\alpha \\ \beta \\ \gamma\end{matrix}\begin{matrix} \\ \\ \\ \delta\ \epsilon\ \eta\end{matrix} = \delta(\alpha,1)\delta(\eta,1)\delta(a^+,b)\delta(f^+,i)\begin{Bmatrix} a & a^+ & 1 \\ d & e & f \\ g & h & f^+ \end{Bmatrix}\begin{matrix}1 \\ \beta \\ \gamma\end{matrix}\begin{matrix} \\ \\ \\ \delta\ \epsilon\ 1\end{matrix} \tag{2.52}$$

with the special case:

$$\begin{Bmatrix} a & a^+ & 1 \\ d & e & f \\ g & h & f^+ \end{Bmatrix}\begin{matrix}1 \\ \beta \\ \gamma\end{matrix}\begin{matrix} \\ \\ \\ \delta\ \epsilon\ 1\end{matrix} = (1/\sqrt{\text{dima}\cdot\text{dimf}})\{h\}\{a^+hee\}\{hf^+g\gamma\}\begin{Bmatrix} d & e & f \\ h & g^+ & a \end{Bmatrix}\begin{matrix} \\ \delta\epsilon\gamma\beta\end{matrix} \tag{2.53}$$

There is an analogue of eq.(2.40) for the 9j symbol, from which we can derive the analogues of (2.41 and 42). The latter is the useful rule of de-Shalit:

$$\sum(\begin{smallmatrix} d & e & f \\ m_d & m_e & m_f \end{smallmatrix})^\beta(\begin{smallmatrix} g & h & i \\ m_g & m_h & m_i \end{smallmatrix})^\gamma(\begin{smallmatrix} a & d & g \\ m_a & m_d & m_g \end{smallmatrix})^\delta(\begin{smallmatrix} b & e & h \\ m_b & m_e & m_h \end{smallmatrix})^\epsilon$$

$$= \sum_{\alpha\eta c}\text{dimc}\cdot\begin{Bmatrix} a & b & c \\ d & e & f \\ g & h & i \end{Bmatrix}\begin{matrix}\alpha \\ \beta \\ \gamma\end{matrix}\begin{matrix} \\ \\ \\ \delta\ \epsilon\ \eta\end{matrix}\cdot(\begin{smallmatrix} a & b & c \\ m_a & m_b & m_c \end{smallmatrix})^\alpha(\begin{smallmatrix} c & f & i \\ m_c & m_f & m_i \end{smallmatrix})^{\eta^*} \tag{2.54}$$

The orthogonality relation of the 9j symbols reads:

$$\sum_{cf}\sum_{\alpha\beta\eta}\text{dimc}\cdot\text{dimf}\cdot\begin{Bmatrix} a & b & c \\ d & e & f \\ g & h & i \end{Bmatrix}\begin{matrix}\alpha \\ \beta \\ \gamma\end{matrix}\begin{matrix} \\ \\ \\ \delta\ \epsilon\ \eta\end{matrix}\cdot\begin{Bmatrix} a & b & c \\ d & e & f \\ g' & h' & i \end{Bmatrix}\begin{matrix}\alpha^* \\ \beta \\ \gamma'\end{matrix}\begin{matrix} \\ \\ \\ \delta'\ e'\ \eta\end{matrix}$$

$$= \delta(h,h')\delta(g,g')\delta(\delta,\delta')\delta(\epsilon,\epsilon')\delta(\gamma,\gamma')/\text{dimg}\cdot\text{dimh} \tag{2.55}$$

Instead of (2.48), the 9j symbol can be defined in terms of 6j symbols:

$$\begin{Bmatrix} a & b & c \\ d & e & f \\ g & h & i \end{Bmatrix}\begin{matrix}\alpha \\ \beta \\ \gamma\end{matrix}\begin{matrix} \\ \\ \\ \delta\ \epsilon\ \eta\end{matrix} = \sum_{\sigma\tau\varphi k}\{k\}\{\text{dimk}\}\begin{Bmatrix} a & b & c \\ h & k & e^+ \end{Bmatrix}_{\sigma\epsilon\varphi\alpha}\begin{Bmatrix} d & e & f \\ k & g^+ & a \end{Bmatrix}_{\delta\sigma\tau\beta}\begin{Bmatrix} g & h & i \\ c^+ & f & k \end{Bmatrix}_{\tau\varphi\eta\gamma} \tag{2.56}$$

Using the orthogonality relation of the 6j symbols, one derives from (2.56) the following sum rule:

$$\sum_{c\alpha\eta} \text{dimc} \cdot \begin{Bmatrix} a & b & c \\ d & e & f \\ g & h & i \\ \delta & \epsilon & \eta \end{Bmatrix} \begin{matrix} \alpha \\ \beta \\ \gamma \end{matrix} \cdot \begin{Bmatrix} a & b & c \\ f^+i & k \end{Bmatrix}_{\mu\pi\eta\alpha} = \{k\} \sum_{\sigma} \begin{Bmatrix} d & e & f \\ b & k & h^+ \end{Bmatrix}_{\sigma\epsilon\pi\beta} \cdot \begin{Bmatrix} g & h & i \\ k & a^+d \end{Bmatrix}_{\delta\sigma\mu\gamma} \quad (2.57)$$

2.7. Tensor operators

By definition, a tensor operator is a set of operators having the transformation property:

$$U(g)T_q^a U(g)^+ = \sum_p D_{pq}^a(g)T_p^a \quad (2.58)$$

The matrix elements of these sets of operators can be factorized according to the Wigner-Eckart theorem (WET):

$$\langle \eta fp | T_q^a | \delta dr \rangle = \sum_\epsilon \langle \eta f \| T^a \| \delta d \rangle_\epsilon \cdot \binom{f^+a\ d}{p\ q\ r}^\epsilon \quad (2.59)$$

From (2.30 and 32) follows the relation of the reduced matrix elements of T^a and of their adjoint operators $(T^a)^+$:

$$\langle \delta d \| (T^a)^+ \| \eta f \rangle_\epsilon = \{f\} \{adf^+\epsilon\} \langle \eta f \| T^a \| \delta d \rangle_\epsilon \quad (2.60)$$

In consequence of (2.34), the coupling of two operators is given by

$$W_m^{ac}(ab) = \{c\} \sqrt{\text{dimc}} \cdot \sum_{kl} \binom{a^+b^+c}{k\ l\ m}^\alpha \cdot U_k^a \cdot V_l^b \quad (2.61)$$

The reduced matrix elements of the set W can be expressed by those of its constituents:

$$\langle \delta d \| W^{ac}(ab) \| \epsilon e \rangle_\beta$$
$$= \{d\} \sqrt{\text{dimc}} \cdot \sum_{\eta f \sigma \tau} \{ad^+f\sigma\} \{ebf^+\tau\} \begin{Bmatrix} a & b & c \\ e & d & f \end{Bmatrix}_{\sigma\tau\beta}^+ \langle \delta d \| U^a \| \eta f \rangle_\sigma \langle \eta f \| V^b \| \epsilon e \rangle_\tau \quad (2.62)$$

If the space of the kets is a direct product space, and if the operators U^a and V^b act on either factor space only, the reduced matrix elements of $W^{ac}(ab)$ can be split up into the reduced matrix elements of U^a and V^b with respect to their factor spaces. Because of (2.34), the states of $H_1 \otimes H_2$ are

$$|(d_1d_2)\delta dr\rangle^{1\otimes 2} = \{d\} \sqrt{\text{dimd}} \cdot \sum \binom{d_1d_2d}{r_1r_2r}^\delta \cdot |d_1r_1\rangle^1 \cdot |d_2r_2\rangle^2, \quad (2.63)$$

and the factorization is given by:

$$\langle\langle (d_1d_2)\delta d \| W^{ac}(ab) \| (e_1e_2)\epsilon e \rangle\rangle_\beta^{1\otimes 2} \quad (2.64)$$
$$= \sqrt{\text{dimd} \cdot \text{dimc} \cdot \text{dime}} \cdot \sum_{\sigma\tau} \begin{Bmatrix} d_1 & d_2 & d^+ \\ a^+ & b^+ & c \\ e_1^+ & e_2^+ & e \end{Bmatrix} \begin{matrix} \delta \\ \alpha \\ \epsilon \end{matrix} \cdot \langle d_1 \| U^a \| e_1 \rangle_\sigma^1 \cdot \langle d_2 \| V^b \| e_2 \rangle_\tau^2$$

2.8. Chains of groups

We now collect the interrelations of functions, coefficients, and reduced matrix elements classified according to a group G' and its subgroup G. The most important case is that of the rotation group and a point group.

In general, a representation a' of G' decomposes into a direct sum of representations a of G,

$$a'(G') = \sum_a n(a',a)a(G) , \qquad (2.65)$$

according to the multiplicity rule:

$$n(a',a) = \sum_{C \in G} \text{ord}C \cdot \chi^a(C)^* \cdot \chi^{a'}(C) \qquad (2.66)$$

The basis functions of $a'(G')$ are adapted to G by the unitary transformation:

$$|a'\alpha a p_a\rangle = \sum \langle a'p'_a | a'\alpha a p_a\rangle \cdot |a'p'_a\rangle \qquad (2.67)$$

$$|a'p'_a\rangle = \sum_{\alpha a} \langle a'\alpha a p_a | a'p'_a\rangle \cdot |a'\alpha a p_a\rangle \qquad (2.68)$$

A definition of the adaption coefficients, which is independent of the special basis, can be given in analogy to (2.22):

$$\text{ord}G^{-1} \sum_{g \in G} D^{a'}_{mn}(g)D^a_{pq}(g)^* = \text{dima}^{-1} \sum_\alpha \langle a'm | a'\alpha a p\rangle \langle a'\alpha a q | a'n\rangle \qquad (2.69)$$

The factor dima^{-1} comes from a different normalization:

$$\sum \langle a'\alpha a p | a'm\rangle \langle a'm | a'\beta b q\rangle = \delta(\alpha,\beta)\delta(a,b)\delta(p,q) \qquad (2.70)$$

$$\sum_{\alpha a} \langle a'm | a'\alpha a p\rangle \langle a'\alpha a p | a'n\rangle = \delta(m,n) \qquad (2.71)$$

All terms classified according to the representations and bases of G' can be adapted to the subgroup G by this transformation. A case of special interest is that of the 3jm symbols:

$$\begin{pmatrix} a' & b' & c' \\ \alpha & \beta & \gamma \\ a & b & c \\ p & q & r \end{pmatrix}^\varepsilon = \sum \begin{pmatrix} a' & b' & c' \\ l & m & n \end{pmatrix}^\varepsilon \langle a'l | a'\alpha a p\rangle \langle b'm | b'\beta b q\rangle \langle c'n | c'\gamma c r\rangle \qquad (2.72)$$

Applying the WET with respect to G to the left hand side, we get:

$$\begin{pmatrix} a' & b' & c' \\ \alpha & \beta & \gamma \\ a & b & c \\ p & q & r \end{pmatrix}^\varepsilon = \sum_\eta \text{Is} \begin{pmatrix} a' & b' & c' \\ \alpha & \beta & \gamma \\ a & b & c \end{pmatrix}^\varepsilon_\eta \cdot \begin{pmatrix} a & b & c \\ p & q & r \end{pmatrix}^\eta . \qquad (2.73)$$

This is nothing but Racah's factorization lemma, which implicitly defines the isoscalar factors (or short isoscalars):

$$\text{Is} \begin{pmatrix} a' & b' & c' \\ \alpha & \beta & \gamma \\ a & b & c \end{pmatrix}^\varepsilon_\eta = \sum \begin{pmatrix} a' & b' & c' \\ l & m & n \end{pmatrix}^\varepsilon \langle a'l | a'\alpha a p\rangle \langle b'm | b'\beta b q\rangle \langle c'n | c'\gamma c r\rangle \begin{pmatrix} a & b & c \\ p & q & r \end{pmatrix}^{\eta *} \qquad (2.74)$$

The isoscalars are invariant to even permutations. Odd permutations yield the phase factors:

$$\text{Is} \begin{pmatrix} a' & b' & c' \\ \alpha & \beta & \gamma \\ a & b & c \end{pmatrix}^\varepsilon_\eta = \{a'b'c'\varepsilon\}\{abc\eta\} \cdot \text{Is} \begin{pmatrix} a' & c' & b' \\ \alpha & \gamma & \beta \\ a & c & b \end{pmatrix}^\varepsilon_\eta \qquad (2.75)$$

The complex conjugation property is:

$$\text{Is} \begin{pmatrix} a'^+ & b'^+ & c'^+ \\ \alpha_+ & \beta_+ & \gamma_+ \\ a^+ & b^+ & c^+ \end{pmatrix}^\varepsilon_\eta = \text{Is} \begin{pmatrix} a' & b' & c' \\ \alpha & \beta & \gamma \\ a & b & c \end{pmatrix}^{\varepsilon *}_\eta \qquad (2.76)$$

The orthogonality relations of the isoscalar factors are:

$$\sum_{\alpha\beta\delta\eta}\text{Is}\begin{vmatrix}a'&b'&c'\\\alpha&\beta&\gamma\\a&b&c\end{vmatrix}_\eta^{\delta^*}\cdot\text{Is}\begin{vmatrix}a'&b'&d'\\\alpha&\beta&\sigma\\a&b&c\end{vmatrix}_\eta^\varepsilon = \delta(c',d')\delta(\delta,\varepsilon)\delta(\gamma,\sigma)\text{dimc/dimc}' \tag{2.77}$$

$$\sum_{\delta\gamma c'}\text{dimc}'\cdot\text{Is}\begin{vmatrix}a'&b'&c'\\\alpha&\beta&\gamma\\a&b&c\end{vmatrix}_\varepsilon^{\delta^*}\cdot\text{Is}\begin{vmatrix}a'&b'&c'\\\sigma&\tau&\gamma\\s&t&c\end{vmatrix}_\eta^\delta = \delta(a,s)\delta(b,t)\delta(\alpha,\sigma)\delta(\beta,\tau)\delta(\varepsilon,\eta)\cdot\text{dimc} \tag{2.78}$$

The isoscalar factor containing the identical representation of G' is very simple:

$$\text{Is}\begin{vmatrix}a'&b'&1'\\\alpha&\beta&\\a&b&1\end{vmatrix}_\eta^\varepsilon = \delta(a'^+,b')\delta(\alpha,\beta)\delta(a^+,b)\delta(\varepsilon,1)\delta(\eta,1)\sqrt{\text{dima/dima}'} \tag{2.79}$$

Further, there is the following sum rule:

$$\sum_{\alpha a}\sqrt{\text{dima}}\cdot\text{Is}\begin{vmatrix}a'&a'^+&c\\\alpha&\alpha_{+\gamma}\\a&a^+1\end{vmatrix}_\eta^\varepsilon = \delta(c',1)\delta(\gamma,1)\delta(\varepsilon,1)\sqrt{\text{dima}'} \tag{2.80}$$

If we invert (2.73) more carefully by using (2.26), we get a more general relation than (2.74):

$$\sum \left(\begin{smallmatrix}a'b'c'\\lmn\end{smallmatrix}\right)^\varepsilon \langle a'l\,|\,a'\alpha ap\rangle\langle b'n\,|\,b'\beta bq\rangle\langle c'n\,|\,c'\gamma cr\rangle\left(\begin{smallmatrix}abd\\pqs\end{smallmatrix}\right)^{\eta^*}$$

$$= \delta(c,d)\delta(r,s)\cdot\text{dimc}^{-1}\cdot\text{Is}\begin{vmatrix}a'&b'&c'\\\alpha&\beta&\gamma\\a&b&c\end{vmatrix}_\eta^\varepsilon \tag{2.81}$$

From the WET and (2.73), we get the relation between the reduced matrix elements with respect to G' and those with respect to its subgroup G:

$$\langle a'\alpha a\|T^{b'\beta b}\|c'\gamma c\rangle_\eta = \sum_\varepsilon \text{Is}\begin{vmatrix}a'^+&b'&c'\\\alpha_{+\beta}&\beta&\gamma\\a^+&b&c\end{vmatrix}_\eta^\varepsilon \cdot \langle a\|T^{b'}\|c\rangle_\varepsilon , \tag{2.82}$$

where the operators, of course, are adapted by:

$$T_p^{b'\beta b} = \sum_m \langle b'm\,|\,b'\beta bp\rangle\cdot T_m^{b'} \tag{2.83}$$

The combination of the WET and group chains has been discussed in [18, 23]. The idea of the isoscalar factor, stimulating the definition of the polyhedral isoscalar in section 6, first came to the authors knowledge by an earlier paper [24], where the isoscalars are termed V-coefficients of the rotation-point group.

2.9. Product groups and double tensors

In the opening of section 2.1., we allowed the symmetry group to be a product group. Since the irreducible representations of a product group are the direct products of the irreducible representations of the single groups, we have a general doubling of the quantum numbers [5], section 3.2:

$$D_{pp',qq'}^{aa'}(gg') = D_{pq}^a(g)\cdot D_{p'q'}^{a'}(g') \tag{2.84}$$

with $g\in G$, $g'\in G'$, and $gg'\in G\times G'$. Consequently the same applies to the characters and the 3jm symbols:

$$\chi^{aa'}(g\tilde{g}) = \chi^{a}(g) \cdot \chi^{a'}(\tilde{g}) \quad , \tag{2.85}$$

$$\begin{pmatrix} aa' & bb' & cc' \\ pp' & qq' & rr' \end{pmatrix}^{\varepsilon\varepsilon'} = \begin{pmatrix} a & b & c \\ p & q & r \end{pmatrix}^{\varepsilon} \cdot \begin{pmatrix} a' & b' & c' \\ p' & q' & r' \end{pmatrix}^{\varepsilon'} \quad , \tag{2.86}$$

and further to all nj symbols. As concernes the states and operators
the doubling applies, too, but in general not the factorization. The
operators with the transformation property

$$U(g\tilde{g})T_{qq'}^{aa'}U(g\tilde{g})^{+} = \sum_{pp'}D_{pq}^{a}(g)D_{p'q'}^{a'}(\tilde{g})T_{pp'}^{aa'} \tag{2.87}$$

are termed double tensor operators. The WET in this case reads:

$$\langle \eta f f p p | T_{qq'}^{aa'} | \delta d d r r \rangle = \sum_{\varepsilon\varepsilon'} \langle \eta f f \| T^{aa'} \| \delta d d \rangle_{\varepsilon\varepsilon'} \begin{pmatrix} f^{+} & a & d \\ p & q & r \end{pmatrix}^{\varepsilon} \begin{pmatrix} f'^{+} & a' & d' \\ p' & q' & r' \end{pmatrix}^{\varepsilon'} \tag{2.88}$$

States and operators allowing a factorization are the (one-particle) spin orbitals and the summands of the spin-orbit coupling operator. On the contrary, the many-elecron states of the ΓS-coupling and
the occupation operators of spin orbitals (cf. section 19) can not be
factorized because of the antisymmetrization involved. Consequently
the coefficients of fractional parentage (cf. section 19) allow no
factorization, too.

The possible factorization is also important to the symmorphic
space groups, since they are direct products of the point and translation groups (cf. section 21).

2.10. The significance of reduced matrix elements

As mentioned in the introduction, the RMEs of one-centre one-particle matrix elements in essence are radial integrals. The same therefore applies to the semi-empirical parameters of the atomic spectroscopy and the ligand field theory even being defined without previous
knowlegde of group theory (cf. the introduction of [25]). In other
cases, the meaning of the RMEs is not so obvious and more indirect:

For compound operators we have the relations (2.62 and 64).

For many-particle states, the RMEs are traced back to the RMEs of
one- and two-particle states by the technique of fractional parentage [26, 27]. We come back to this subject in section 19 by a different approach.

The RMEs of the symmetry-adapted LCAO-functions have not been ana-
lysed, so far, and are the main issue of this treaty (cf. section 5
and in a more general context section 20).

3. Representations induced by polyhedral edges

This section closely follows [9, 11]. In the theory of molecules, the first step towards quantitative considerations of symmetry is the formation of symmetry-adapted (s.-a. in the following) linear combinations, especially molecular orbitals and symmetric coordinates. This is well known and done heuristically in simple cases or systematically by the projection operator technique. The insertion of such symmetrized molecular orbitals into molecular matrix-elements results in rather complex expressions [28]. We therefore prefer another approach by studying the symmetry-adaption of arbitrary "objects" defined with respect to the vertices and edges of the polyhedra (and later on with respect to the faces, too).

We consider molecules $A_m B_n C_p \ldots$ with a symmetry group G. This may be optionally a point group, a double point group, or the direct product of a point group and the spin group SU(2).

The positions of equivalent atoms A, B etc., which constitute a symmetric polyhedron, are indicated by the vectors $\vec{A}_i$, $\vec{B}_k$ etc, where i and k give the numbering within the equivalent sets. The distance vectors $\vec{S}_{ik} = \vec{A}_i - \vec{A}_k$, $\vec{T}_{ik} = \vec{B}_i - \vec{B}_k$, $\vec{U}_{ik} = \vec{A}_i - \vec{B}_k$ etc. likewise form equivalent sets. In order to simplify the notation and to allow differences of the distance vectors again, we use beside the double indices a simple numbering $\vec{S}_{ik} \rightarrow \vec{S}_m$. The correspondence of $\vec{S}_m$ with $\vec{A}_i$ and $\vec{A}_k$ or of $\vec{U}_r$ with $\vec{A}_i$ and $\vec{B}_k$ can be expressed by a topological matrix $\tau\binom{ABU}{ikr}$, which disapears, if not $\vec{A}_i + \vec{B}_k + \vec{U}_r = 0$. The value in the other case could be chosen equal to one, but, because of the analogy with the 3jm symbols, it is more appropriate to choose $1/\sqrt{Z(ABU)}$, where Z(ABU) is the number of all triangles equivalent to $\vec{A}_i + \vec{B}_k + \vec{U}_r = 0$. The analogy mentioned will be treated in section 6. We sum up the definition:

$$\tau\binom{ABU}{ikr} = \begin{cases} 1/\sqrt{Z(ABU)} & \text{for } \vec{A}_i + \vec{B}_k + \vec{U}_r = 0 \\ 0 & \text{for } \vec{A}_i + \vec{B}_k + \vec{U}_r \neq 0 \end{cases} \tag{3.1}$$

Now the correspondence $\vec{U}_{ik} \rightarrow \vec{U}_r$ can be expressed as follows:

$$\vec{A}_i - \vec{B}_k = \sum_{Um} \sqrt{Z(-ABU)} \cdot \tau\binom{-ABU}{ikm} \cdot \vec{U}_m \tag{3.2}$$

Later on we shall use correspondences of this type also for other numbered objects in the polyhedral framework.

The simple numbering allows to treat the equivalent sets S on an equal footing with the sets A of atomic positions. We thus can write in a unified manner the (in general reducible) representation σ^S of the symmetry group G, induced by the equivalent set S:

$$g\vec{S}_i = \sum_k \sigma^S_{ki}(g)\vec{S}_k \tag{3.3}$$

According to the character formula (2.10) this representation σ^S car be decomposed into a direct sum (2.11). For finite and compact groups, this decomposition is guaranteed by Maschke's theorem [29]. Moreover, it is proven that the decomposition is achieved by a unitary transformation [30]:

$$\sum_{ik}(S\alpha ar|\vec{S_i})\sigma^S_{ik}(g)(\vec{S_k}|S\beta bs) = \delta(a,b)\delta(\alpha,\beta)D^a_{rs}(g) \tag{3.4}$$

or

$$\sum_{k}\sigma^S_{ik}(g)(\vec{S_k}|S\alpha as) = \sum_{r}D^a_{rs}(g)(\vec{S_i}|S\alpha ar) \tag{3.5}$$

with

$$\sum_{i}(S\alpha ar|\vec{S_i})(\vec{S_i}|S\beta bs) = \delta(a,b)\delta(\alpha,\beta)\delta(r,s) \tag{3.6}$$

and

$$\sum_{\alpha ar}(\vec{S_i}|S\alpha ar)(S\alpha ar|\vec{S_k}) = \delta(i,k) , \tag{3.7}$$

where α and β are the multiplicity indices. On analogy of (2.69), the transformation coefficients also can be defined by:

$$\text{ordG}^{-1}\sum_{g\in G}\sigma^S_{ik}(g)D^a_{rs}(g)^* = \text{dima}^{-1}\sum_{\alpha}(\vec{S_i}|S\alpha ar)(S\alpha as|\vec{S_k}) \tag{3.8}$$

This relation is derived from (3.5) by (3.7) and (2.4). The notation of the transformation matrix is chosen following Dirac s bra-ket formalism. The analogy is twofold. At first $(\vec{S_i}|S\alpha ap)$ is analogous to $\langle am|a\alpha ap\rangle$. In the same way as we adapt functions to the group G by (2.67), we can adapt s-functions, $s(\vec{r})=\langle\vec{r}|s\rangle$, centered at the atomic positions $\vec{A_i}$ by:

$$\langle\vec{r}|sA\alpha ap\rangle = \sum_{i}\langle\vec{r}-\vec{A_i}|s\rangle(\vec{A_i}|A\alpha ap) \tag{3.9}$$

This is the formation of s.-a. LCAO-MOs by linear combination. Quoting Cotton [31], we therefore call the coefficients $(\vec{A_i}|A\alpha ap)$ SALC coefficients (i.e. coefficients of symmetry-adapted linear combinations). More details concerning this aspect are elaborated in section 5. We can invert (3.9) by (3.7):

$$\langle\vec{r}-\vec{A_i}|s\rangle = \sum_{\alpha ap}\langle\vec{r}|sA\alpha ap\rangle(A\alpha ap|\vec{A_i}) \tag{3.10}$$

which, of course, is analogous to (2.68).

But there is another interpretation of (3.10). $\langle\vec{r}-\vec{A_i}|s\rangle = s(\vec{r}-\vec{A_i})$ may be regarded as a function of the discrete variable A_i. (3.10) then gives an expansion of these functions into $(A\alpha ap|\vec{A_i})$ with expansion coefficients $\langle\vec{r}|sA\alpha ap\rangle$ according to (3.9). Thus the SALC coefficients turn out to be s.-a. functions of the positions $\vec{A_i}$. From this point of view they are analogous to s.-a. Hilbbert space functions r!φap . Because this second analogy is very important and fruitful, we work it out in six parallel steps:

1a) Consider the Euclidean space R_3.
1b) Consider the set $A = \{\vec{A_1}, \vec{A_2},...\vec{A_{Z(A)}}\}$, where $Z(A)$ is the number of equivalent positions in A.

2a) Define the functions $\varphi(\vec{r})=\langle\vec{r}|\varphi\rangle$ on R_3, i.e. $\vec{r}\in R_3$.

2b) Define the functions $G(\vec{A_i})=(\vec{A_i}|G)$ on A, i.e. $\vec{A_i}\in A$.[+]

3a) The kets $|\varphi\rangle$ subtend a Hilbert space $H(R_3)$, $\langle\vec{r}|\varphi\rangle$ being the space representation of $|\varphi\rangle$.

3b) The kets $|G)$ subtend a finite, unitary space $U(A)$, $(\vec{A_i}|G)$ being the space representation of $|G)$.

4a) In $H(R_3)$ there are s.-a. functions $\langle\vec{r}|\varphi ap\rangle$ transforming according to (2.17).

4b) In $U(A)$ there are s.-a. functions $(\vec{A_i}|A\alpha ap)$ transforming according to

$$(g^{-1}\vec{A_i}|A\alpha ap) = \sum D^a_{qp}(g)(\vec{A_i}|A\alpha aq) \qquad (3.11)$$

Because of (3.3) the transformation property (3.11) is identical with (3.5). This means, that the s.-a. functions in $U(A)$ are just the SALC coefficients, if the appropriate normalization is chosen.

5a) The s.-a. functions $\langle\vec{r}|\varphi ap\rangle$ are orthogonal and complete, if

$$\sum_{\varphi ap}\langle\vec{r}|\varphi ap\rangle\langle\varphi ap|\vec{r}'\rangle = \delta(\vec{r},\vec{r}') \qquad (3.12)$$

$$\int\langle\varphi ap|\vec{r}\rangle\langle\vec{r}|\varphi ap''\rangle d^3 r = \delta(a,a')\delta(\varphi,\varphi')\delta(p,p') \qquad (3.13)$$

5b) The s.-a. functions $(\vec{A_i}|A\alpha ap)$ are already orthogonal and complete because of (3.6/7).

6a) Because of (3.12/13), we can expand every function $\langle\vec{r}|\eta\rangle\in H(R_3)$ according to

$$\langle\vec{r}|\eta\rangle = \sum_{\varphi ap}\langle\varphi ap|\eta\rangle\langle\vec{r}|\varphi ap\rangle \qquad (3.14)$$

with the expansion coefficients:

$$\langle\varphi ap|\eta\rangle = \int\langle\varphi ap|\vec{r}\rangle\langle\vec{r}|\eta\rangle d^3 r \qquad (3.15)$$

6b) Because of (3.6/7), we can expand every function $(\vec{A_i}|G)\in U(A)$ according to

$$(\vec{A_i}|G) = \sum_{\alpha ap}(A\alpha ap|G)(\vec{A_i}|A\alpha ap) \qquad (3.16)$$

with the expansion coefficients:

$$(A\alpha ap|G) = \sum_i(A\alpha ap|\vec{A_i})(\vec{A_i}|G) . \qquad (3.17)$$

This expansion theorem, which is based on (3.6/7) is the main result of our discussion. Eq. (3.10), which was our starting point, is now

---

[+] Here a note on our phraseology may be in order. In a puristic way of speaking, $\sin(x)$ is no function, but a value of the function sin. In the same sense $G(\vec{A_i})$ or $(\vec{A_i}|G)$ is no function (or vector in $U(A)$), but a value of the function (or a component of the vector) $|G)$. But to simplify matters, we keep calling $(\vec{A_i}|G)$ a vector, sinx a function, and a_{ik} a matrix.

a special case of (3.16). The symmetry-adaption (3.9) is a special case of (3.17). Therefore in general (Aαap|G) may be interpreted as the s.-a. version of $(\vec{A}_i|G)$.

The functions $(\vec{A}_i|G) \in U(A)$ often result from the ordinary function of $H(R_3)$ by inserting edge vectors, i.e. $\vec{r}=\vec{A}_i$. A case of special interest are the spherical harmonics

$$\langle\vec{r}|lm\rangle = i^l Y_{lm}(\vec{r}/r) , \tag{3.18}$$

where the phase is chosen in accordance to section 2.3, i.e. $\langle\vec{r}|l^+m\rangle= \langle lm|\vec{r}\rangle$, and the s.-a. spherical harmonics:

$$\langle\vec{r}|l\alpha ap\rangle = \sum\langle lm|l\alpha ap\rangle\langle\vec{r}|lm\rangle \tag{3.19}$$

If we insert the argument $\vec{A}_i$ into these functions, (3.16/17) yield:

$$\langle\vec{A}_i|l\alpha ap\rangle = 0 \text{ for } n(A,a) = 0, \tag{3.20}$$

otherwise;

$$\langle\vec{A}_i|l\alpha ap\rangle = \sum_{\beta=1}^{n(A,a)} (\vec{A}_i|A\beta ap)\cdot c(A\beta a,l\alpha) \tag{3.21}$$

$$c(A\beta a,l\alpha) = \sum_i (A\beta ap|\vec{A}_i)\langle\vec{A}_i|l\alpha ap\rangle \text{ (no } \sum_p !) \tag{3.22}$$

At the first sight, this important result is somewhat surprising. As functions over R_3 there are infinitely many, linearly independent spherical harmonics, but as functions over the set A the most of them are linearly dependent. This results from the special directions enforced upon the edge vectors by the symmetry. The expansion coefficients (3.22) do not depend on the numbering of edges and the choice of coordinate axes. They are a property of the polyhedral framework. The expansion (3.21) has been used in [32] to expand the spherical harmonics of intergral formulae. A table of the coefficients for a tetrahedral molecule is also given there (cf. also table 26).

If the representation a is contained only once in σ^A, (3.21) provides us with a simple tool to calculate the SALC coefficients. We only have to insert $\vec{A}_i$ into a s.-a. spherical harmonic tabulated in [21] and get a SALC coefficient up to a normalization factor. This method has first been used in [10]. If the multiplicity exceeds one, problems of linear independence and orthogonality arise and the coefficients are determined only up to a unitary transformation:

$$(\vec{A}_i|A\alpha ap) = \sum_\beta u_{\alpha\beta}^{Aa}\cdot(\vec{A}_i|A\beta ap) \tag{3.23}$$

For each set A and each irreducible representation a, a new choice has to be made. These problems are settled systematically in section 12, where also a uniform choice of SALC coefficients will be proposed for all edges of all molecules sharing one symmetry group. In the same section the calculation of the expansion coefficients (3.22) is reduced to a smaller class of coefficients.

Because of the one-to-one correspondence $\vec{A}_i$, $\vec{B}_k \to \vec{U}_r$ according to (3.2), every function $f(\vec{A}_i,\vec{B}_k)$ can be regarded as well as a function of the related $\vec{U}_r$:

$$f(\vec{A}_i,\vec{B}_k) = \sum_{\vec{U}m} \sqrt{Z(-ABU)}\tau\binom{-ABU}{ikm}f(\vec{U}_m) = f(\vec{U}_r) \qquad (3.24)$$

If the funtions considered depend on the difference $\vec{A}_i-\vec{B}_k$ only, the special case $\vec{B}_k=\vec{A}_i$ makes no trouble. The edge vector $\vec{U}_r$ is simply equal to the zero vector. This natural point of view has been adopted in the papers [10, 12, 32]. But if the functions depend on $\vec{A}_i$ and $\vec{B}_k$ separately, we need a more sophisticated treatment of the case $\vec{B}_k=\vec{A}_i$. We have to discriminate the zero edge vectors $\vec{A}_i-\vec{A}_i=\vec{0}_i^A$ according to their position (i.e. with respect to set A and number i), since the functions $f(\vec{0}_i^A)$ do depend on A and i.

This new interpretation is consistent with the following correspondence of edge vectors between atoms, i.e. $\vec{S}_{ik}=\vec{A}_i-\vec{B}_k$, and of vectors centred in the origin. These vectors are defined by:

$$\vec{S}_{ik}' = \mu_1\vec{A}_i + \mu_2\vec{B}_k \qquad (3.25)$$

An incommensurable choice of the coefficients μ_1 guarantees the a one-to-one mapping, i.e. $\vec{S}_{ik}'\neq\vec{T}_{lm}'$ and $\vec{S}_{ik}'\neq\vec{S}_{ki}'$ especially. Because of this mapping $\vec{S}_{ik} \to \vec{S}_{ik}'$, all functions of $\vec{S}_{ik}$ can be regarded as functions of $\vec{S}_{ik}'$. Especially holds:

$$(\vec{S}_{ik}|S\alpha ap) = (\vec{S}_{ik}'|S'\alpha ap) \qquad (3.26)$$

In the case of zero edge vectors we have $\vec{0}_i^A= \mu_1\vec{A}_i+\mu_2\vec{A}_i=(\mu_1+\mu_2)\vec{A}_i$, which yields the reasonable correspondence $\vec{0}_i^A \to \vec{A}_i$. By (3.18), the symmetry-adaption with respect of arbitrary polyhedral edges is reduced to the symmetry-adaption with respect to vectors in the centre of symmetry. Correspondences of this type will prove to be useful also for objects related to more than two centres.

4. Bicentric matrices

4.1. Molecular integrals

The central theorem of the quantitative group theory is that of
Wigner and Eckart. It applies to matrix elements of s.-a. functions
and tensor operators refering to the same centre of symmetry. On the
contrary, all calculations of molecules starting from localized orbitales at the atomic positions lead to polycentric matrix elements.
The following theorem applies to bicentric matrix elements of tensor
operators and includes the WET as a special case.

We proceed from an arbitrary set of s.-a. functions of species a:

$$\varphi_p^a(\vec{r}) = \langle \vec{r}|\varphi ap\rangle, \tag{4.1}$$

i.e. with the tranformational property according to (2.6/17):

$$\langle \vec{r}|U(g)|\varphi ap\rangle = \langle g^{-1}\vec{r}|\varphi ap\rangle = \sum D_{qp}^a(g)\langle \vec{r}|\varphi aq\rangle \tag{4.2}$$

By translation to the different atomic positions, we generate the
orbitals:

$$\langle \vec{r}|Ai\varphi ap\rangle = \langle \vec{r}-\vec{A_i}|\varphi ap\rangle \tag{4.3}$$

As usual in quantum chemistry [33], the position is not indicated by
a vector, but the set A and the number i are treated as additional
quantum numbers. The special case of s-orbitals has already been mentioned in (3.9).

The tensor operators are defined by (2.58). In what follows, the operators refer to the centre of symmetry or are invariant under translations, as the operators of momentum and kinetic energy. The general case
of shifted operators is treated separately in section 8. The main example of this case are the potential operators. But we stress that s.-a.
sums, like $V = \sum_i f(|\vec{r}-\vec{A_i}|)$, are allowed for, because the entire sum re-
fers to the centre of symmetry. The difference between these sums
and the invariant operators does not matter in the present discussion,
but will give rise to the distinction of proper and improper bicentric
matrix elements later on.

After these preliminaries, we state the factorization theorem for
the bicentric matrix elements:

$$\langle Ai\varphi_a am|T_n^c|Bk\varphi_b bp\rangle$$

$$= \sum_{d\delta\eta\varepsilon e}(A\varphi_a a\|T^c\|B\varphi_b b)_{S\varepsilon e}^{d\delta\eta}\cdot\begin{pmatrix}a^+b & d\\ m & p & q\end{pmatrix}^\delta\begin{pmatrix}d^+c & e\\ q & n & r\end{pmatrix}^\eta\cdot(\vec{S}_{ik}|S\varepsilon er) \tag{4.4}$$

with $\vec{S}_{ik} = \vec{A_i}-\vec{B_k}$. If we skip all discriminating multiplicity indices,
the geometric structure contrasts better:

$$\langle Aiam|T_n^c|Bkbp\rangle = \sum_{de}(Aa\|T^c\|Bb)_{Se}^d\cdot\begin{pmatrix}a^+b & d\\ m & p & q\end{pmatrix}\begin{pmatrix}d^+c & e\\ q & n & r\end{pmatrix}\cdot(S_{ik}|Ser) \tag{4.5}$$

The invariants $(A\varphi_a a\|T^C\|B\varphi_b b)^{d\delta\eta}_{S\epsilon e}$ defined by (4.4) are independent of the numbering of atoms and the choice of axes. In general they depend on the lengths A, B, and S.

Taking into account (3.2),(4.4) can be given another form, which at the first sight looks more complicated. But it proves to be useful for the further calculations, because the topological matrix connecting $\vec{A_i}$, $\vec{B_k}$, and $\vec{S_r}$ leads to important, geometric results.

$$\langle Ai\varphi_a am|T^C_n|Bk\varphi_b bp\rangle \qquad (4.6)$$

$$= \sum_{d\delta\eta\epsilon e}\sum_{St}(A\varphi_a a\|T^C\|B\varphi_b b)^{d\delta\eta}_{S\epsilon e}\cdot\binom{a^+b\ \ d}{m\ \ p\ \ q}\delta\binom{d^+c\ \ e^+}{q\ \ n\ \ r}\eta\cdot\tau\binom{-ABS}{ikt}\sqrt{Z(-ABS)}(\vec{S_t}|S\epsilon er)$$

The sum for S runs over all sets of equivalent edges and the topological matrix selects the right one. This is necessary, since the vertices of the tetrahedron or the octahedron are connected to each other by inequivalent sets of edges.

Proof of (4.4): In the case of a translationally invariant operator, the matrix elements are functions of the distance vectors $\vec{S}_{ik}$ only. In the case of an operator related to the centre of symmetry, they are moreover functions of A and B, because the triangles are specified definitely by A, B, and $\vec{S}_{ik}$. In both cases, they are functions of the edge vectors $\vec{S}_{ik}$ and therefore expandable according to theorem (3.16):

$$\langle Ai\varphi_a am|T^C_n|Bk\varphi_b bp\rangle = \sum_{\epsilon e}(S\epsilon er|X)(\vec{S}_{ik}|S\epsilon er) \qquad (4.7)$$

with the collective index $X=(Ai\varphi_a amTcnBk\varphi_b bp)$. According to (3.17) the expansion coefficients are

$$(S\epsilon er|X) = \sum_{ik}(S\epsilon er|S_{ik})\langle Ai\varphi_a am|T^C_n|Bk\varphi_b bp\rangle , \qquad (4.8)$$

where the sum for i, k is limited to $\vec{A_i}-\vec{B_k}=\vec{S}_{ik}$ (i.e. the distance must belong to set S). The coefficients (4.8) transform as a direct product $a^+b\times c\times e^+$ and thus can be factorized. For this purpose, we only have to generalize the WET to a fourfold direct product:

$$(S\epsilon er|X) = \sum_{d\delta\eta}(A\varphi_a a\|T^C\|B\varphi_b b)^{d\delta\eta}_{S\epsilon e}\cdot\binom{a^+b\ \ d}{m\ \ p\ \ q}\delta\binom{d^+c\ \ e^+}{q\ \ n\ \ r}\eta \qquad (4.9)$$

If we insert (4.9) into (4.7), the proof is complete.

From (4.8/9), we can isolate the invariant and express it by all matrix elements:

$$(A\varphi_a a\|T^C\|B\varphi_b b)^{d\delta\eta}_{S\epsilon e}$$

$$= \{b\}\dim d\cdot\sum_{ik}\binom{a\ \ b^+d}{m\ \ p\ \ q}\delta\binom{d\ \ c^+e}{q\ \ n\ \ r}\eta(S\epsilon er|S_{ik})\langle Ai\varphi_a am|T^C_n|Bk\varphi_b bp\rangle \qquad (4.10)$$

with the same limitation of i, k as in (4.8). We have used the phase $\{e\}=1$, because e must be a tensor representation even for double point groups, i.e. including spin-orbitals in the matrix elements.

If integral formulae are at hand, we can calculate the invariants by (4.10). Since the wit of (4.4) is to express many matrix elements by few invariants, this calculation is only efficient, if the 3jm symbols and SALC coefficients subsequently can be eliminated. In section 15 such a calculation is carried out. The invariants defined by (4.4) or (4.10) shall be called BRM (bicentric, reduced matrix element).

Because of the relevance of scalar operators, we repeat this special case of (4.4/6):

$$\langle Ai\varphi_a am|T|Bk\varphi_b bp \rangle = \sum_{\delta\varepsilon e} (A\varphi_a a\|T\|B\varphi_b b)^{\delta}_{S\varepsilon e} \cdot \binom{a^+ b \ e^+}{m \ p \ r}^{\delta} dime^{-1/2} (\vec{S}_{ik}|S\varepsilon er) \quad (4.11)$$

$$= \sum_{\delta\varepsilon e} \sum_{St} \sqrt{Z(-ABS)/dime} \cdot (A\varphi_a a\|T\|B\varphi_b b)^{\delta}_{S\varepsilon e} \binom{a^+ b \ e^+}{m \ p \ r}^{\delta} \tau \binom{-ABS}{ikt} (\vec{S}_t|S\varepsilon er)$$

where more precisely $(A\varphi_a a\|T\|B\varphi_b b)^{\delta}_{S\varepsilon e} = (A\varphi_a a\|T\|B\varphi_b b)^{e^+\delta 1}_{S\varepsilon e}$.

4.2. General bicentric matrices

The relevance of (4.11) goes far beyond the one-particle integrals. By careful inspection of the proof of (4.4) we observe that the matrix does not need to be an integral. It only has to transform in the right way under the symmetry operations. Thus we can reshape (4.11) as a general matrix theorem:

In the following $\vec{X}_r$ are arbitrary vectors fixed in a molecular framework (i.e. not necessarily distance vectors between atoms). If we can show that a bicentric matrix $M(\vec{X}_r)_{Ai\varphi_a am, Bk\varphi_b bp}$ transforms according to

$$M(g^{-1}\vec{X}_r)_{Ai\varphi_a am, Bk\varphi_b bp} \quad (4.12)$$
$$= \sum_{JI} \sum_{nq} D^a_{nm}(g)\sigma^A_{ji}(g) D^b_{qp}(g)\sigma^B_{lk}(g) \cdot M(\vec{X}_r)_{Aj\varphi_a an, Bl\varphi_b bp} \ )$$

then the bicentric matrix has the following factorization:

$$M(\vec{X}_r)_{Ai\varphi_a am, Bk\varphi_b bp} = \sum_{\delta\varepsilon e} M(A\varphi_a a\|B\varphi_b b)^{\delta}_{S\varepsilon e} \binom{a^+ b \ e^+}{m \ p \ r}^{\delta} dime^{-1/2} (\vec{S}_{ik}|S\varepsilon er) \quad (4.13)$$

The generalized BRM $M(A\varphi_a a\|B\varphi_b b)^{\delta}_{S\varepsilon e}$ is a scalar function of the vectors $\vec{X}_r$ with respect to the symmetry group G. In the present context, φ_a and φ_b mean additional indices not affected by the symmetry operations.

Examples of such bicentric matrices being no integrals are the inverse overlap matrix of the atomic orbitals in a molecule, the one-particle density matrix in the AO basis, and the matrix of the vibrational force constants with respect to the atomic elongations. The last example has already been dealt with in [12]. The analogous generalization of (4.4) would be needed, if one considers anharmonic effects.

The inversion of (4.13) is:

$$M(A\varphi_a a \| B\varphi_b b)^{\delta}_{S\varepsilon e} = \{b\}\sqrt{\text{dime}} \cdot \sum_{ik} \binom{a\ b^+ e}{m\ p\ r}^{\delta}(S\varepsilon er|S_{ik})M(\vec{X}_r)_{Ai\varphi_a am, Bk\varphi_b bp} \quad (4.14)$$

As a first, simple example, we now can determine the BRMs of the unit matrix in the AO basis: $E_{Ai\varphi_a am, Bk\varphi_b bp} = \delta(A,B)\delta(i,k)\delta(\varphi_a,\varphi_b)\delta(a,b)\delta(m,p)$
The result is:

$$E(A\varphi_a a \| B\varphi_b b)^{\delta}_{S\varepsilon e}$$
$$= \delta(A,B)\delta(\varphi_a,\varphi_b)\delta(a,b)\delta(S,O^A)\delta(e,1)\delta(\varepsilon,1)\delta(\delta,1)\sqrt{Z(A)\cdot\text{dime}} \quad (4.15)$$

Since the product of two bicentric matrices is a bicentric matric again, the BRMs of the product matrix is determined by those of the factors. But we postpone this calculation, because we need some results of section 6.

The integral and non-integral, bicentric matrices often refer not directly to the irreducible representations of the molecular symmetry group G, but to the angular momentum basis, for the atomic orbitals and elongations are at first given in this basis. We then are confronted with matrices of the type $M(\vec{X}_r)_{Ain_a l_a m_a, Bkn_b l_b m_b}$. Since the angular momentum basis is in general reducible with respect to G, we first have to adapt the basis according to (3.19) and then apply (4.13). This yields the general structure of bicentric matrices in the angular momentum basis:

$$M(\vec{X}_r)_{Ain_a l_a m_a, Bkn_b l_b m_b} = \sum_{\alpha\alpha\beta b}\sum_{\delta\varepsilon e}M(An_a l_a \alpha a \| Bn_b l_b \beta b)^{\delta}_{S\varepsilon e}\langle l_a m_a | l_a \alpha a p_a\rangle \quad (4.16)$$
$$\cdot \langle l_b \beta b p_b | l_b m_b\rangle \binom{a^+ b\ e}{p_a p_b p_e}^{\delta}\cdot\text{dime}^{-1/2}(\vec{S}_{ik}|S\varepsilon e p_e)$$

5. The matrix elements of s.-a. molecular orbitals

The s.-a. MOs can be built up from the AOs $|Ai\varphi_a ap_a\rangle$ defined by (4.1/3).
As has been shown in [10, 11], the complete set of symmetrized LCAOs
resulting from an equivalent set $|Ai\varphi_a ap_a\rangle$ is given by:

$$|(A\varepsilon e, \varphi_a a)\gamma cp_c\rangle = \sum_{ip_a} K(\gamma cp_c, Ai\varepsilon e, ap_a) \cdot |Ai\varphi_a ap_a\rangle \qquad (5.1)$$

with the compound SALC coefficient:

$$K(\gamma cp_c, Ai\varepsilon e, ap_a) = \{c\}\sqrt{dimc} \cdot \sum \binom{e^+ a^+ c}{p_e p_a p_c}^\gamma \cdot (\overrightarrow{A_i}|A\varepsilon ep_e) \qquad (5.2)$$

In comparison to [10, 11], the phase factor $\{c\}$ has been included in
order to allow for spin-orbitals. The definition is in accordance
with (2.34).

The proof of (5.1/2) results from the transformation property of
the AOs:

$$U(g)|Ai\varphi_a ap\rangle = \sum_{kq} D^a_{qp}(g)\sigma^A_{ki}(g) \cdot |Ak\varphi_a aq\rangle \qquad (5.3)$$

Since the SALC coefficients reduce the representation σ^A and the 3jm
symbols the remaining product representation, the compound SALC coefficients reduce the direct product $a \times \sigma^A$ in (5.3).

In the appendix 1 this method of symmetry-adaption is compared to
the conventional technique of the projection operators.

If we now consider the matrix elements of tensor operators, they
can be factorized as usual by the WET:

$$\langle\!\langle(A\varepsilon e, \varphi_a a)\gamma cp_c|T^g_{pg}|(B\varphi f, \varphi_b b)\delta dp_d\rangle$$
$$= \sum_\alpha \langle\!\langle(A\varepsilon e, \varphi_a a)\gamma c\|T^g\|(B\varphi f, \varphi_b b)\delta d\rangle_\alpha \binom{c^+ g \ d}{p_c p_g p_d}^\alpha \qquad (5.4)$$

Now the ordinary, reduced matrix element of (5.4) is related to the
BRM in the following theorem:

$$\langle\!\langle(A\varepsilon e, \varphi_a a)\gamma c\|T^g\|(B\varphi f, \varphi_b b)\delta d\rangle_\alpha = \{d\}\{c^+ dg\alpha\}\sqrt{dimc \cdot dimd}$$

$$\cdot \sum_{h\eta\theta\beta S\sigma k}\sqrt{Z(-ABS)}\begin{Bmatrix} c^+ d & g \\ e & f^+ k^+ \\ a & b^+ h^+ \end{Bmatrix}^\alpha_\beta \cdot PIs\begin{pmatrix} -A & B & S \\ \varepsilon_+ \varphi & \sigma \\ e^+ f & k \end{pmatrix}^\theta_\eta \cdot (A\varphi_a a\|T^g\|B\varphi_b b)^{h\eta\theta}_{S\sigma k} \qquad (5.5)$$

In this theorem appears a typical, geometric invariant of the polyhedral framework, the "polyhedral isoscalar". It refers to the equivalent triangles (-ABS) and is defined by (6.6). The details are discussed in the following section.

The derivation of the theorem (5.5) is as follows. We first solve
(5.4) for the reduced matrix element and then insert (5.1) into the
ordinary matrix elements. This yields sums over bicentric matrix elements, which are replaced by (4.6). The five resulting 3jm symbols are

collected in a 9j and one 3jm symbol using the rule of de-Shalit (2.54).
Except for the BRM and the 9j symbol the remainder is equal to the
right side of (6.6) and thus suggests the definition of the polyhedral isoscalar.

The skipping of all additional indices in (5.5) shows again the essential structure, which is given by a triple sum only:

$$\langle\!\langle Ae,a)c\|T^g\|(Bf,b)d\rangle = \{d\}\{c^+dg\}\sqrt{dimc\cdot dimd}$$
$$\cdot\sum_{hSk}\sqrt{Z(-ABS)}\cdot\left\{\begin{matrix}c^+d&g_+\\e&f^+k^+\\a&b^+h^+\end{matrix}\right\}\cdot PIs\left(\begin{matrix}-A&B&S\\e^+f&k\end{matrix}\right)\cdot(Aa\|T^g\|Bb)^h_{Sk} \tag{5.6}$$

If we regard the reduced matrix elements of the s.-a. MOs as the
physical properties of a molecule, we can say: The theorem (5.5/6) expresses the non-local invariants representing the physical properties
by the BRMs, the local invariants of the coordination. This connection
is mediated by the sums for S and k, i.e. by a sum over the different
coordinations of the atoms and a sum over the representations k contained in σ^S. The sum for h is more technical. As d in the sum (4.5)
it counts the multiplicity of the identical representation in the fourfold product $a^+k b \times g \times k^+$ (cf. also eq.(7.15)).

If T^g is the Hamiltonian of the system at hand, the BRMs (with $S\neq0$)
on the right side are the invariant representatives of the resonance
integrals and describe the chemical binding along the polyhedral edges S. These invariants thus are the appropriate candidates for a
semi-empirical parametrization (Hückel, Wolfsberg-Helmholtz). The author is of the opinion that relevant parameters must not depend on
the numbering of atoms and the choice of axes (cf. section 13).

From this interpretation we conclude that (5.5) is the first, polyhedral example of the structure (1.2). The geometrical factor in this
case is

$$GEO_1(x,y,z) = \{d\}\{c^+dg\alpha\}\sqrt{Z(-ABS)dimc\cdot dimd}\cdot\sum_\beta\left\{\begin{matrix}c^+d&g_+&\alpha\\e&f^+k^+&\beta\\a&b^+h^+&\eta\\\gamma&\delta&\theta\end{matrix}\right\}\cdot PIs\left(\begin{matrix}-A&B&S\\\varepsilon&\varphi&\sigma\\e^+f&k\end{matrix}\right|_\beta \tag{5.7}$$

with $x=(AaBbg)$, $y=(\varepsilon e\gamma c,\eta f\delta d,\alpha)$, and $z=(h\eta\theta S\sigma k)$. (5.5) then reads:

$$\langle\!\langle A\varepsilon e,\varphi_a a)\gamma c\|T^g\|(B\varphi f,\varphi_b b)\delta d\rangle_\alpha = \sum_z GEO_1(x,y,z)\cdot(A\varphi_a a\|T^g\|B\varphi_b b)^{h\eta\theta}_{S\sigma k} \tag{5.8}$$

In order to invert this relation we need an orthogonality relation
of the geometrical factors for a fixed constellation $x=(AaBbg)$. From
the relations (2.55) and (6.9/10) we derive:

$$\sum_y GEO_1(x,y,z)^*\cdot GEO_1(x,y,z') = \delta(z,z')/dimh \tag{5.9}$$

$$\sum_z dimh\cdot GEO_1(x,y,z)^*\cdot GEO_1(x,y',z) = \delta(y,y') \tag{5.10}$$

and further the inversion of (5.8):

$$(A\varphi_a a \| T^g \| B\varphi_b b)_{S\sigma k}^{h\eta\theta} = \sum_y dimh \cdot GEO_1(x,y,z) \cdot \langle\!\langle A\varepsilon e,\varphi_a a)\gamma c \| T^g \| (B\varphi f,\varphi_b b)\delta d \rangle\!\rangle_\alpha$$

(5.11)

This relation may be used to adjust semi-empirical parameters.

As in the preceeding section, we write the formulae for the scalar operators again separately, whereby the 9j symbol reduces according to (2.52/53). (5.4) then becomes:

$$\langle\!\langle A\varepsilon e,\varphi_a a)\gamma c p_c | T | (B\varphi f,\varphi_b b)\delta d p_d \rangle\!\rangle$$

(5.12)

$$= \delta(c,d)\delta(p_c,p_d)\sqrt{dimc}\,\langle\!\langle A\varepsilon e,\varphi_a a)\gamma c \| T \| (B\varphi f,\varphi_b b)\delta d \rangle\!\rangle$$

and (5.5):

$$\langle\!\langle A\varepsilon e,\varphi_a a)\gamma c \| T \| (B\varphi f,\varphi_b b)\delta d \rangle\!\rangle = \delta(c,d)\{b\}\{c^+ f b \delta\}\sqrt{dimc}$$

(5.13)

$$\cdot \sum_{\eta\beta}\sum_{S\sigma k}\{ab^+ k\eta\}\sqrt{Z(-ABS)/dimk}\cdot\left\{\begin{matrix}e & f^+ & k^+ \\ b^+ & a^+ & c^+\end{matrix}\right\}_{\gamma\delta\eta\beta} PIs\left(\begin{matrix}-A & B & S \\ \varepsilon & \varphi & \sigma \\ e^+ f & k\end{matrix}\right)_\beta \cdot (A\varphi_a a \| T \| B\varphi_b b)_{S\sigma k}^\eta$$

6. Polyhedral coefficients referring to edges

In section 3, we have noted the parallelism of the SALC coefficients $(\vec{A_i}|A\alpha ar)$ and the subgroup-adaption coefficients $\langle \acute{a}m|\acute{a}\alpha ar \rangle$, since both types of coefficients decompose reducible representations of the group G. The analogous pairs of equations are (2.69) and (3.8), (2.70) and (3.6), (2.71) and (3.7). We now pursue this parallelism further and show that it also comprises the 3jm symbols of G' and the isoscalar factors of $G' \supset G$.

The representations $\acute{a}(\acute{G})$ etc. are coupled by the 3jm symbols of the group G', which in this section is supposed to be simply reducible. In other words the 3jm symbols of G' couple reducible representations of G. This observation suggests that there might be also a coupling of the reducible representations σ^S of G. Indeed, this is true and the part of the coupling coefficients is now played by the topological matrices $\tau(^{ABC}_{ikl})$. The analogue of eq. (2.25) is:

$$\sum_{ilp}\sigma^A_{ik}(g)\sigma^B_{lm}(g)\sigma^C_{pq}(g)\tau(^{ABC}_{ilp}) = \tau(^{ABC}_{kmq}) \tag{6.1}$$

This equation represents the mapping of the triangle $\vec{A_i}+\vec{B_l}+\vec{C_p}=0$ onto the equivalent triangle $\vec{A_k}+\vec{B_m}+\vec{C_q}=0$ by the operation g. The analogue of the orthogonality relation (2.26) is:

$$\sum_{kl}\tau(^{ABC}_{klm})\tau(^{ABD}_{kln}) = \delta(C,D)\delta(m,n)/Z(C) , \tag{6.2}$$

where Z(C) is the number of edge vectors of type C. Note the possible difference of Z(ABC) and Z(C), since one edge vector of type C may be shared by different triangles of type ABC. In this case, C must be invariant to some symmetry operations. Each $\vec{C_1}$ is shared by $q=Z(ABC)/Z(C)$ triangles. Since $\vec{A_k}+\vec{B_l}+\vec{C_m}=0$ excludes $\vec{A_k}+\vec{B_l}+\vec{D_n}=0$, the left side of (6.2) is zero unless $\vec{C_m}=\vec{D_n}$. If now $\vec{C_m}=\vec{D_n}$, there are q non-zero summands and, because of the normalization chosen in (3.1), eq.(6.2) is valid.

If the vectors of type A and B point from the centre of symmetry to certain positions, we have Z(ABC)=Z(C) and each $\vec{C_1}$ uniquely determines a pair $(\vec{A_i}, \vec{B_k})$. Only in this case, we get the second orthogonality condition,

$$\sum_{Cr}Z(C)\tau(^{ABC}_{ikr})\tau(^{ABC}_{lmr}) = \delta(i,l)\delta(k,m) , \tag{6.3}$$

the analogue of (2.27). When using consequences of (6.3), we allways have to make shure that Z(ABC)=Z(C).

We now proceed as with the 3jm symbols of the supergroup G and transform the topological matrix into the s.-a. basis (cf.(2.72)):

$$\tau\begin{pmatrix} A & B & C \\ \alpha & \beta & \gamma \\ a & b & c \\ k & l & m \end{pmatrix} = \sum_{rst}\tau(^{A\ B\ C}_{r\ s\ t})(\vec{A_r}|A\alpha ak)(\vec{B_s}|B\beta bl)(\vec{C_t}|C\gamma cm) \tag{6.4}$$

According to the WET, we get the analogue of Racah's factorization lemma (2.73),

$$\tau \begin{pmatrix} A & B & C \\ \alpha & \beta & \gamma \\ a & b & c \\ k & l & m \end{pmatrix} = \sum_{\varepsilon} PIs \begin{pmatrix} ABC \\ \alpha\beta\gamma \\ abc \end{pmatrix}_{\varepsilon} \cdot \binom{abc}{klm}^{\varepsilon} \ , \tag{6.5}$$

which defines the "polyhedral isoscalar factor" (cf.(2.74)):

$$PIs \begin{pmatrix} ABC \\ \alpha\beta\gamma \\ abc \end{pmatrix}_{\varepsilon} = \sum_{rst} \tau\binom{ABC}{rst}(\vec{A}_r|A\alpha ak)(\vec{B}_s|B\beta bl)(\vec{C}_t|C\gamma cm)\binom{abc}{klm}^{\varepsilon} \tag{6.6}$$

Since the topological matrices are real and (by choice) invariant to
all permutations of columns, we have the following symmetries of the
polyhedral isoscalars. They are invariant to even permutations and
odd permutations yield a phase factor:

$$PIs \begin{pmatrix} ABC \\ \alpha\beta\gamma \\ abc \end{pmatrix}_{\varepsilon} = \{abc\varepsilon\} \cdot PIs \begin{pmatrix} ACB \\ \alpha\gamma\beta \\ acb \end{pmatrix}_{\varepsilon} \tag{6.7}$$

The complex conjugation is simple:

$$PIs \begin{pmatrix} ABC \\ \alpha\beta\gamma \\ abc \end{pmatrix}_{\varepsilon}^{*} = PIs \begin{pmatrix} A & B & C \\ \alpha_+ & \beta_+ & \gamma_+ \\ a^+ & b^+ & c^+ \end{pmatrix}_{\varepsilon} \tag{6.8}$$

From (6.2) we derive the first orthogonality relation of the polyhedral
isoscalars (cf.(2.77)):

$$\sum_{\alpha\alpha\beta b\varepsilon} PIs \begin{pmatrix} ABC \\ \alpha\beta\gamma \\ abc \end{pmatrix}_{\varepsilon}^{*} PIs \begin{pmatrix} ABD \\ \alpha\beta\sigma \\ abc \end{pmatrix}_{\varepsilon} = \delta(C,D)\delta(\gamma,\sigma)\text{dim}c/Z(C) \tag{6.9}$$

The analogue of (2.78) following from (6.3) is subjected to the restriction Z(ABC)=Z(C):

$$\sum_{C\gamma} Z(C)PIs \begin{pmatrix} ABC \\ \alpha\beta\gamma \\ abc \end{pmatrix}_{\varepsilon}^{*} PIs \begin{pmatrix} ABC \\ \sigma\tau\gamma \\ stc \end{pmatrix}_{\eta} = \delta(a,s)\delta(b,t)\delta(\alpha,\sigma)\delta(\beta,\tau)\delta(\varepsilon,\eta)\text{dim}c \tag{6.10}$$

Analogous to (2.79/80), we derive the special case

$$PIs \begin{pmatrix} ABO \\ \alpha\beta \\ ab1 \end{pmatrix}_{\varepsilon} = \delta(-A,B)\delta(\alpha,\beta)\delta(a^+,b)\delta(\varepsilon,1)\sqrt{\text{dim}a}/Z(A), \tag{6.11}$$

and the sum rule:

$$\sum_{\alpha a} \sqrt{\text{dim}a} \cdot PIs \begin{pmatrix} -A & A & C \\ \alpha & \alpha_+\gamma \\ a & a^+1 \end{pmatrix} = \delta(C,0^A)\delta(\gamma,1) \tag{6.12}$$

Instead of (6.6), we can derive from (6.5), by the help of (2.26), the
more general relation analogous to (2.81):

$$\sum_{rst} \tau\binom{ABC}{rst}(\vec{A}_r|A\alpha ak)(\vec{B}_s|B\beta bl)(\vec{C}_t|C\gamma cm)\binom{abd}{kln}^{\varepsilon}$$

$$= \delta c,d)\delta(m,n) \cdot \text{dim}c^{-1} \cdot PIs \begin{pmatrix} ABC \\ \alpha\beta\gamma \\ abc \end{pmatrix}_{\varepsilon} \tag{6.13}$$

This relation allows several rearrangements like:

$$\sum_{rs} \tau \binom{ABC}{rst} (\vec{A_r}|A\alpha ak)(\vec{B_s}|B\beta bl) = \sum_{\epsilon\gamma c} PIs \begin{vmatrix} ABC \\ \alpha\beta\gamma \\ abc \end{vmatrix}_\epsilon \binom{abc}{klm}^\epsilon (C\gamma cm|\vec{C_t}) \qquad (6.14)$$

A first application of the polyhedral isoscalars we have already met in (5.5) because we got an expression including the right hand side of (6.6). The polyhedral isoscalars are the group theoretical, i.e. invariant representation of the triangular conditions like $\vec{A_i}+\vec{B_k}+\vec{C_1}=0$. As we shall see later on, such triple relations of "polyhedral objects" can be generalized.

Another application, already announced in section 4, results from the product of two bicentric matrices:

$$M_{A i \varphi_a am, Bk\varphi_b bp} = \sum_{Cl\varphi_c cn} \sum P_{A i \varphi_a am, Cl\varphi_c cn} \cdot Q_{Cl\varphi_c cn, Bk\varphi_b bp} \qquad (6.15)$$

Inserting this into (4.14) yields the BRM of the product matrix in terms of the BRMs of the factors. The derivation resembles that of (5.5) and results in:

$$M(A\varphi_a \| B\varphi_b b)^\delta_{S\epsilon e} = \{b\}\{ab^+e\delta\} \sum_{C\varphi_c c} \sum_{T\tau\eta f U\tau\eta f} \sum \frac{\sqrt{Z(STU)dime}}{\sqrt{dimf \cdot dimf'}} \cdot \begin{Bmatrix} e & f^+ f' \\ c & b & a \end{Bmatrix}_{\delta\tau\tau'\alpha}$$

$$\cdot PIs \begin{vmatrix} S & T & U \\ \epsilon+f & \eta & \eta' \\ e+f & f' \end{vmatrix}_\alpha \cdot P(A\varphi_a a \| C\varphi_c c)^\tau_{T\eta f} Q(C\varphi_c c \| B\varphi_b b)^{\tau'}_{U\eta' f'} \qquad (6.16)$$

7. Triangular coefficients

7.1. Representations induced by triangles

In the same way as the polyhedral edges, the triangles and distorted tetrahedra subtended in the molecular framework form equivalent sets with respect to the symmetry group. These sets again carry reducible representations of the group. This is of interest for the functions depending on the triangles or pseudo-tetrahedra, for instance the molecular three- or four-centre integrals. With regard to this application it is necessary to distinguish the triangles and pseudo-tetrahedra by valued or numbered vertices. This distinction accords to our treatment of the edges as vectors, i.e. as line segments with orientation or valued ends. The numbering becomes essential, if the the vertices of the triangle or pseudo-tetrahedron are equivalent to each other. Thus a triangle or pseudo-tetrahedro is invariant to a symmetry operation only, if all its vertices lie on the reflection plane or the rotation axis. As for the edges we have to take into account degenerate cases, i.e. triangles and pseudo-tetrahedra with coincident vertices. The extreme cases are the null-triangles and null-tetrahedra with three or four coinciding vertices.

In this section we discuss the symmetry-adaption of triangles. If Λ is a set of equivalent triangles Δ_i, then these triangles carry the representation σ^Δ analogous to (3.3):

$$g\Delta_i = \sum_k \sigma^\Delta_{ki}(g)\Delta_k \tag{7.1}$$

with the characters

$$\sigma^\Delta(g) = \sum_i \sigma^\Delta_{ii}(g) \tag{7.2}$$

They are equal to the number of triangles invariant to the operation g. Again the representation σ^Δ is decomposed according to the branching rule (2.10/11) with a multiplicity $n(\Delta,a)$. In analogy to (3.4/5), the unitary transformation is given by:

$$\sum_{ik}(\Delta\alpha ar|\Delta_i)\sigma^\Delta_{ik}(g)(\Delta_k|\Delta\beta bs) = \delta(a,b)\delta(\alpha,\beta)D^a_{rs}(g) \tag{7.3}$$

$$\sum_k \sigma^\Delta_{ik}(g)(\Delta_k|\Delta\beta bs) = \sum_r D^a_{rs}(g)(\Delta_i|\Delta\alpha ar) \tag{7.4}$$

The matrix elements $(\Delta_i|\Delta\alpha ar)$ are termed triangular SALC (TSALC) coefficients. The unitary relations analogous to (3.6/7) are:

$$\sum_i (\Delta\alpha ar|\Delta_i)(\Delta_i|\Delta\beta bs) = \delta(a,b)\delta(\alpha,\beta)\delta(r,s) \tag{7.5}$$

$$\sum_{\alpha ar}(\Delta_i|\Delta\alpha ar)(\Delta\alpha ar|\Delta_k) = \delta(i,k) \tag{7.6}$$

Up to now we have used an arbitrary numbering of the triangles

irrespective of the numbers of their vertices. The interrelation of
the vertices and triangles is again expressed by a topological matrix:

$$\tau\left(\begin{smallmatrix}\Delta ACB \\ ikml\end{smallmatrix}\right) = \begin{cases} 1/\sqrt{Z(\Delta)} & \text{if the ordered triple } \vec{A_k}, \vec{C_m}, \vec{B_l} \text{ represents the} \\ & \text{vertices of } \Delta_i \quad (7.7) \\ 0 & \text{otherwise}, \end{cases}$$

where $Z(\Delta)$ is the number of equivalent triangles in the set Δ. The order ACB has been chosen with regard to the molecular three-centre integrals (cf.(8.1/3)).

The topological matrix is utilized to reduce the triangular SALC
coefficients $(\Delta_i|\Delta\alpha ar)$ to the ordinary SALC coefficients. The technique
is similar to (3.25/26). Each triangle Δ_j is mapped onto a vector $\vec{R_j}$:

$$\vec{R_j} = \sum_{kml}\tau\left(\begin{smallmatrix}\Delta ACB \\ jkml\end{smallmatrix}\right)(\mu_1\vec{A_k}+\mu_2\vec{C_m}+\mu_3\vec{B_l}) \tag{7.8}$$

By the choice of the fixed numbers μ_i, we have to take care that $R_j \neq R_k$
if $\Delta_j \neq \Delta_k$. Since this mapping is bijective, we have $\sigma^R = \sigma^\Delta$. Therefore
the triangular SALC coefficients of the set Δ are equal to the ordina-
ry SALC coefficients of the set R:

$$(\Delta_j|\Delta\alpha ar) = (\vec{R_j}|R\alpha ar) \tag{7.9}$$

In principle, this relation would allow the total elimination of the
triangles Δ by the vectors R. Because of (7.8), an integral with the
three centres $\vec{A_k}$, $\vec{C_m}$, and $\vec{B_l}$ can be regarded as a function of one vector $\vec{R_j}$. Although this interpretation is valid, it might be confusing.
So we keep to the more expressive notation Δ_j and use (7.9) only for
the calculation of $(\Delta_j|\Delta\alpha ar)$.

Using (7.8), we explicitly show that any function of three positions can be regarded as a function over a set of triangles:

$$F(\Delta_i) = (\Delta_i|F) = \sum_{kml}\tau\left(\begin{smallmatrix}\Delta ACB \\ ikml\end{smallmatrix}\right) \cdot F(A_k, C_m, B_l) \tag{7.10}$$

The notation $(\Delta_i|F)$ suggests the same proceeding as in section 3 and
the interpretation of (7.5/6) as the orthogonality and completeness
relations of a set of s.-a. triangle functions. This argument leads
to the conclusion that every triangle function $(\Delta_i|F)$ can be expanded
analogous to (3.16/17):

$$(\Delta_i|F) = \sum_{\alpha a}(\Delta_i|\Delta\alpha ar)(\Delta\alpha ar|F) \tag{7.11}$$

with the expansion coefficients:

$$(\Delta\alpha ar|F) = \sum_i(\Delta\alpha ar|\Delta_i)(\Delta_i|G) \tag{7.12}$$

If such a function depends on the intrinsic parameters of the triangles
only, but not on their orientation in space, then it is invariant on
the set Δ. In this case, we simply write $F(\Delta)$ or F_Δ. An example is the
reduced matrix element in eq.(8.1).

7.2. The coupling of triangular representations

The topological matrices (7.8) are quite analogous to (3.1) and there-fore have corresponding orthogonality relations:

$$\sum_{kml} \tau(^{\Delta ACB}_{ikml})\tau(^{\theta ACB}_{jkml}) = \delta(\Delta,\theta)\delta(i,j)/Z(\Delta) \qquad (7.13)$$

$$\sum_{\Delta i} Z(\Delta)\tau(^{\Delta ACB}_{ikml})\tau(^{\Delta ACB}_{inpq}) = \delta(k,n)\delta(m,p)\delta(l,q) \qquad (7.14)$$

Since Δ_i uniquely determines the three vertices $\vec{A}_k$, $\vec{B}_l$, and $\vec{C}_m$ (7.14) is not subjected to a restriction as (6.3).

The topological matrices (7.8) have four columns of indices. A di-rect treatment in the sense of section 6 would lead to a parallelism with the group-theoretical 4jm symbols, which couple four irreducible representations [34]. But the 4jm symbols can be factorized into 3jm symbols in the following way:

$$(^{abcd}_{ijkl})\epsilon\eta e = \sum_m (^{abe}_{ijm})\epsilon(^{e^+cd}_{m\ kl})\eta_{dime} \qquad (7.15)$$

The collective index $\epsilon\eta e$ counts the multiplicity of the identical re-prsentation in the direct product $a \times b \times c \times d$. This suggests to try a corresponding decomposition of the topological matrix (7.8), which is achieved as follows. Instead by its three vertices, a triangle can be determined by one of its edges and the opposite vertex. The interrela-tion of the triangle Δ_i, the edge $\vec{S}_k$, and the vertex $\vec{C}_l$ is expressed by a further topological matrix. For the purpose of discrimination from (3.1), we call it a topological matrix of second kind:

$$\tau^2(^{\Delta SC}_{irm}) = \begin{cases} 1/\sqrt{Z(\Delta)} & \text{if } \vec{C}_m \text{ is the 2nd vertex and } \vec{S}_r \text{ points from the} \\ & \qquad\qquad \text{1st to the 3rd vertex of } \Delta_i \quad (7.16) \\ 0 & \text{otherwise} \end{cases}$$

The original interrelation (7.8) now can be decomposed:

$$\tau(^{\Delta ACB}_{ikml}) = \sum_{Sr} \tau^2(^{\Delta SC}_{irm})\tau(^{-ABS}_{klr})\sqrt{Z(-ABS)} \qquad (7.17)$$

The orthogonality relations follow immediately from the definition:

$$\sum_{rm} \tau^2(^{\Delta SC}_{irm})\tau^2(^{\theta SC}_{jrm}) = \delta(\Delta,\theta)\delta(i,j)/Z(\Delta) \qquad (7.18)$$

$$\sum_{\Delta i} Z(\Delta)\tau^2(^{\Delta SC}_{irm})\tau^2(^{\Delta SC}_{ipq}) = \delta(r,q)\delta(m,p) \qquad (7.19)$$

Of course, (7.18/19) combined with (6.2/3) yield (7.13/14). Again there is no restriction for (7.19), since Δ_i uniquely determines $\vec{S}_r$ and $\vec{C}_m$.

We now strictly follow the proceeding in section 6 from eq.(6.4) onwards. The transformation of τ^2 into the s.-a. basis yields:

$$\tau^2 \begin{pmatrix} \Delta SC \\ \delta\sigma\gamma \\ dsc \\ pqr \end{pmatrix} = \sum_{ikm} \tau^2(^{\Delta SC}_{ikm})(\Delta_i|\Delta\delta dp)(\vec{S}_k|So sq)(\vec{C}_m|C\gamma cr) \qquad (7.20)$$

and again the factorization by the WET:

$$\tau^2 \begin{pmatrix} \Delta SC \\ \delta\sigma\gamma \\ dsc \\ pqr \end{pmatrix} = \sum_\varepsilon PIs^2 \begin{pmatrix} \Delta SC \\ \delta\sigma\gamma \\ dsc \end{pmatrix}_\varepsilon \begin{pmatrix} dsc \\ pqr \end{pmatrix}^\varepsilon \tag{7.21}$$

with the "polyhedral isoscalar of the second kind":

$$PIs^2 \begin{pmatrix} \Delta SC \\ \delta\sigma\gamma \\ dsc \end{pmatrix}_\varepsilon = \sum_{\overline{ikm}} \tau^2 \begin{pmatrix} \Delta SC \\ ikm \end{pmatrix} (\Delta_i | \Delta\delta dp)(\vec{S}_k | S\sigma sq)(\vec{C}_m | C\gamma cr) \begin{pmatrix} dsc \\ pqr \end{pmatrix}^{\varepsilon*} \tag{7.22}$$

The orthogonality relations follow from (7.18/19):

$$\sum_{\sigma s} \sum_{\gamma c \varepsilon} PIs^2 \begin{pmatrix} \Delta SC \\ \delta\sigma\gamma \\ dsc \end{pmatrix}_\varepsilon^* PIs^2 \begin{pmatrix} \Theta SC \\ \vartheta\sigma\gamma \\ dsc \end{pmatrix}_\varepsilon = \delta(\Delta,\Theta)\delta(\delta,\vartheta) \text{dimd}/Z(\Delta) \tag{7.23}$$

$$\sum_{\Delta\delta} Z(\Delta) PIs^2 \begin{pmatrix} \Delta SC \\ \delta\sigma\gamma \\ dsc \end{pmatrix}_\varepsilon^* PIs^2 \begin{pmatrix} \Delta SC \\ \delta\sigma\gamma' \\ dsc' \end{pmatrix}_\eta = \delta(\sigma,\sigma')\delta(s,s')\delta(\gamma,\gamma')\delta(c,c')\delta(\varepsilon,\eta)\cdot\text{dimd} \tag{7.24}$$

(7.2) again can be generalized to

$$\sum_{\overline{jkm}} \tau^2 \begin{pmatrix} \Delta SC \\ jkm \end{pmatrix} (\Delta_j | \Delta\delta dp)(\vec{S}_k | S\sigma sq)(\vec{C}_m | C\gamma cr) \begin{pmatrix} dsc \\ pqr \end{pmatrix}^{\varepsilon*}$$
$$= \delta(s,s')\delta(q,q')\text{dims}^{-1} PIs^2 \begin{pmatrix} \Delta SC \\ \delta\sigma\gamma \\ dsc \end{pmatrix}_\varepsilon \tag{7.25}$$

This can be derived from (7.21) by the orthonality relation (2.26).
The relations (2.40), (2.81), (6.14), and (7.25) are all of the same
type and can be proved by a uniform method. We have a term of the form
$X\begin{pmatrix} ss' \\ qq' \end{pmatrix}$, the left hand side of these equations. Starting now from the
expression $\sum_q D^s_{pq}(g) X\begin{pmatrix} ss' \\ qq' \end{pmatrix}$ we shift the operation g in X from term to term.
In the case of (7.25) for instance, we use the relations

$$\sum_i (\Delta\alpha ar | \Delta_i) \sigma_{ik}(g) = \sum_s D^a_{rs}(g)(\Delta\alpha as | \Delta_k) , \tag{7.26}$$

$$\sum_{km} \sigma^S_{lk}(g) \sigma^C_{nm}(g) \tau^2 \begin{pmatrix} \Delta SC \\ jkm \end{pmatrix} = \sum_i \tau^2 \begin{pmatrix} \Delta SC \\ iln \end{pmatrix} \sigma_{ij}(g) , \tag{7.27}$$

and (2.25). This shifting finally results in:

$$\sum_q D^s_{pq}(g) X\begin{pmatrix} ss' \\ qq' \end{pmatrix} = \sum_r X\begin{pmatrix} ss' \\ pr \end{pmatrix} D^{s'}_{rq}(g) \tag{7.28}$$

Thus X would achieve a similarity transformation between irreducible
representations. According to Schur's lemma follows:

$$X\begin{pmatrix} ss' \\ qq' \end{pmatrix} = \delta(s,s')\delta(q,q')\cdot X\begin{pmatrix} ss \\ qq \end{pmatrix} ,$$

where $X\begin{pmatrix} ss \\ qq \end{pmatrix}$ does not depend on q. By summation for q with (7.22), finally follows (7.25).

8. Triangular invariants

The most important examples of functions defined on triangles are the molecular three-centre nuclear-attraction integrals. Because of the expansion theorem (7.11) they can be factorized as follows:

$$\langle A i \varphi_a am \mid \mid \vec{r}-\vec{C_1}\mid^{-1}\mid Bk\varphi_b bq\rangle \tag{8.1}$$

$$= \sum_{\varepsilon\gamma c}\sum_{\Delta j}(A\varphi_a\|Cr^{-1}\|B\varphi_b b)^{\varepsilon}_{\Delta\gamma c}\sqrt{Z(\Delta)}\tau\binom{\Delta ACB}{jikl}\binom{a^+b\ c}{m\ q\ n}^{+\varepsilon}(\Delta_j\mid\Delta\gamma cn)$$

By this relation the tricentric, reduced matrix elements $(A.a\|C.\|B.b)^{\varepsilon}_{\Delta\gamma c}$ are defined, the abbreviation of which shall be TRM. Because of the limited, graphical possibilities the symbol of the TRM turns out somewhat similar to that of the BRM. But the third centre of the operator and the index referring to a set of equivalent triangles should suffice for a distinction.

In order to prove (8.1), we consider the expression,

$$I^{\varepsilon c}_n(\vec{A_1},\vec{C_1},\vec{B_k}) = \sum\binom{a^+b\ c}{m\ q\ n}^{+\varepsilon*}\cdot\text{dimc}\cdot\langle A i\varphi_a am\mid\mid\vec{r}-\vec{C_1}\mid^{-1}\mid Bk\varphi_b bq\rangle. \tag{8.2}$$

According to (7.10), $I^{\varepsilon c}_n$ is a function of the triangles Δ_j:

$$I^{\varepsilon c}_n(\Delta_j) = \sum_{ilk}\sqrt{Z(\Delta)}\tau\binom{\Delta ACB}{jilk}\cdot I^{\varepsilon c}_n(\vec{A_1},\vec{C_1},\vec{B_k}) \tag{8.3}$$

Noting (7.11), we can readily expand $I^{\varepsilon c}_n$ in triangular SALC coefficients. Since $I^{\varepsilon c}_n$ transforms according to the representation c, the expansion is limited to the coefficients $(\Delta_j\mid\Delta\gamma cn)$:

$$I^{\varepsilon c}_n(\Delta_j) = \sum_{\gamma}(A\varphi_a a\|Cr^{-1}\|B\varphi_b b)^{\varepsilon}_{\Delta\gamma c}(\Delta_j\mid\Delta\gamma cn) \tag{8.4}$$

We now invert the relations (8.2) and (8.3), using (2.27) and (7.14):

$$\langle A i\varphi_a am\mid\mid\vec{r}-\vec{C_1}\mid^{-1}\mid Bk\varphi_b bq\rangle = \sum_{\varepsilon cn}\binom{a^+b\ c}{m\ q\ n}^{\varepsilon}\cdot I^{\varepsilon c}_n(\vec{A_1},\vec{C_1},\vec{B_k}) \tag{8.5}$$

$$I^{\varepsilon c}_n(\vec{A_1},\vec{C_1},\vec{B_k}) = \sum_{\Delta j}\sqrt{Z(\Delta)}\tau\binom{\Delta ACB}{jilk}\cdot I^{\varepsilon c}_n(\Delta_j) \tag{8.6}$$

Inserting now (8.4) and (8.6) into (8.5) we get (8.1).

If we invert (8.4) by (7.5), we get

$$(A\varphi_a a\|Cr^{-1}\|B\varphi_b b)^{\varepsilon}_{\Delta\gamma c} \tag{8.7}$$

$$= \sum_{jilk}(\Delta\gamma cn\mid\Delta_j)\sqrt{Z(\Delta)}\tau\binom{\Delta ACB}{jilk}\text{dimc}\binom{a^+b\ c}{m\ q\ n}^{\varepsilon*}\langle A i\varphi_a am\mid\mid\vec{r}-\vec{C_1}\mid^{-1}\mid B\varphi_b bq\rangle$$

which is needed in section 15.

Because of the factorization of the topological matrix given in (7.17), eq.(8.1) takes the following form, which is appropriate for further calculations:

$$\langle A i\varphi_a am\mid\mid\vec{r}-\vec{C_1}\mid^{-1}\mid Bk\varphi_b bq\rangle = \sum_{\varepsilon\gamma c\Delta jSr}\sum(A\varphi_a a\|Cr^{-1}\|B\varphi_b b)^{\varepsilon}_{\Delta\gamma c}\sqrt{Z(\Delta)Z(-ABS)'}$$

$$\cdot\tau\binom{-ABS}{ikr}\tau^2\binom{\Delta SC}{jrl}\binom{a^+b\ c}{m\ q\ n}^{\varepsilon}(\Delta_j\mid\Delta\gamma cn) \tag{8.8}$$

The main application of (8.1 or 8) is the calculation of the matrix elements of the nuclear potential

$$V = \sum_{Cl} Q_C / |\vec{r} - \vec{C}_1| \, , \tag{8.9}$$

where Q_C is the charge of the atoms of set C. The potential contains the totally symmetric partial sums

$$V_C = \sum_{l} 1 / |\vec{r} - \vec{C}_1| \, , \tag{8.10}$$

because these sums are just the totally symmetric linear combinations of the operators $1/|\vec{r} - \vec{C}_1|$. We rewrite these sums in the terminology of (5.1/2). The symmetry species occurring in these formulae are all totally symmetric, i.e. a=e=c=1:

$$V_C = \sqrt{Z(C)} \cdot \sum_{l} (\vec{C}_1 | C11) \cdot |\vec{r} - \vec{C}_1|^{-1} \tag{8.11}$$

Since the improper two-centre integrals of the entire molecular potential trace back to the three-centre nuclear-attraction integrals, we now can expresses the BRMs of the molecular potential V, or ratherof the partial sums V_C by the TRMs. Because of (8.11), the matrix elements are

$$\langle A i \varphi_a am | V_C | B k \varphi_b bq \rangle = \sqrt{Z(C)} \sum_{l} (\vec{C}_1 | C11) \langle A i \varphi_a am | \, |\vec{r} - \vec{C}_1|^{-1} | B k \varphi_b bq \rangle$$

and further with (8.8):

$$\langle A i \varphi_a am | V_C | B k \varphi_b bq \rangle = \sum_{\varepsilon \gamma c} \sum_{\Delta S r} (A \varphi_a a \| Cr^{-1} \| B \varphi_b b)^{\varepsilon}_{\Delta \gamma c} \sqrt{Z(\Delta) Z(-ABS) Z(C)}$$

$$\cdot \tau \binom{-ABS}{ikr} \binom{a^+ b \ c^+}{m \ q \ n}^{\varepsilon} \sum_{Jl} \tau^2 \binom{\Delta SC}{jrl} (\Delta_j | \Delta \gamma cn)(\vec{C}_1 | C11) \tag{8.12}$$

Reshaping (7.25), the sum for j and l is found to be

$$\sum_{Jl} \tau^2 \binom{\Delta SC}{jrl} (\Delta_j | \Delta \gamma cn)(\vec{C}_1 | C11) = dimc^{-1} \sum_{\sigma} PIs^2 \begin{pmatrix} \Delta & S & C \\ \gamma & \sigma & \sigma \\ c & c^+1 \end{pmatrix} (\vec{S}_r | S\sigma cn) . \tag{8.13}$$

If we now insert (8.13) into (8.12) and then introduce the result into (4.10), we finally get the intended relationship:

$$(A \varphi_a a \| V_C \| B \varphi_b b)^{\varepsilon}_{S\sigma c} = \sum_{\Delta \gamma} (A \varphi_a a \| Cr^{-1} \| B \varphi_b b)^{\varepsilon}_{\Delta \gamma c} \sqrt{Z(\Delta) Z(C)} \cdot PIs^2 \begin{pmatrix} \Delta & S & C \\ \gamma & \sigma & \sigma \\ c & c^+1 \end{pmatrix} \tag{8.14}$$

This shows that all the TRMs, the orbitals of which adjoin to the edges of type S, contribute to the BRMs belonging to type S. The geometric relationship is mediated by a polyhedral isoscalar of the second kind, which is relatively simple:

$$PIs^2 \begin{pmatrix} \Delta & S & C \\ \gamma & \sigma & \sigma \\ c & c^+1 \end{pmatrix} = (1/\sqrt{Z(C)dimc}) \sum_{Jrl} \tau^2 \binom{\Delta SC}{jrl} (\Delta_j | \Delta \gamma cq)(S\sigma cq | \vec{S}_r) \tag{8.15}$$

Introducing (8.14) into (5.13) allows to express all matrix elements of the potential V_C with respect to the s.-a. MOs by the TRMs. Since this relationship is often needed, we integrate it by a geometrical factor:

$$\langle\!\langle (A\epsilon e, \varphi_a a)\gamma c\|V_c\|(B\varphi f, \varphi_b b)\delta\dot{c}\rangle\!\rangle = \sum_z GEO_2(x,y,z)\cdot(A\varphi_a a\|Cr^{-1}\|B\varphi_b b)^{\eta}_{\Delta\alpha k} \qquad (8.16)$$

with the collective indices $x=(Aa,Bb,C)$, $y=(\epsilon e,\varphi f,\gamma\delta c)$ and $z=(\Delta\alpha k,\eta)$.
The geometrical factor is given by.

$$GEO_2(x,y,z) = \sqrt{Z(\Delta)Z(C)dimc}\cdot\sum_{\beta S\sigma k}\sqrt{Z(-ABS)/dimk}\{b\}\{c^+fb\delta\}\{eb^+k\eta\}$$

$$\cdot\begin{Bmatrix}e & f^+k^+ \\ b^+a^+c^+\end{Bmatrix}_{\gamma\delta\eta\beta}PIs\begin{pmatrix}-A & B & S \\ \epsilon & \varphi & \sigma \\ e^+f & k\end{pmatrix}_{\beta}PIs^2\begin{pmatrix}\Delta & S & C \\ \alpha & \sigma \\ k & k^+\imath\end{pmatrix} \qquad (8.17)$$

The double sum for S and σ connecting the two polyhedral isoscalars results from the decomposition (7.17) and represents the higher polyhedral
isoscalar related to (7.7):

$$PIs\begin{pmatrix}\Delta & A & C & B \\ \alpha & \epsilon & & \varphi \\ k & e^+\imath & f & \beta\end{pmatrix} = \sum_{S\sigma}\sqrt{Z(-ABS)}\cdot PIs\begin{pmatrix}-A & B & S \\ \epsilon & \varphi & \sigma \\ e^+f & k\end{pmatrix}_{\beta}\cdot PIs^2\begin{pmatrix}\Delta & S & C \\ \alpha & \sigma \\ k & k^+\imath\end{pmatrix} \qquad (8.18)$$

$$= \sqrt{1/Z(C)}\sum_{j\imath lk}\tau\begin{pmatrix}\Delta ACB \\ j\imath lk\end{pmatrix}(A\epsilon ep|\vec{A}_\imath)(\vec{B}_k|B\varphi fq)(\Delta_j|\Delta\alpha kr)\begin{pmatrix}e^+f & k \\ p & q & r\end{pmatrix}\beta^*$$

If this isoscalar is introduced, the geometrical factor takes the form:

$$GEO_2(x,y,z) = \sqrt{Z(\Delta)Z(C)}\sum_{\beta k}\sqrt{dimc/dimk}\{b\}\{c^+fb\delta\}\{eb^+k\eta\}$$

$$\cdot\begin{Bmatrix}e & f^+k^+ \\ b^+a^+c^+\end{Bmatrix}_{\gamma\delta\eta\beta}\cdot PIs\begin{pmatrix}\Delta & A & C & B \\ \alpha & \epsilon & & \varphi \\ k & e^+\imath & f & \beta\end{pmatrix} \qquad (8.19)$$

9. Symmetry-adapted geminals and densities

Preparatory to the discussion of the two-particle integrals, we explain the composition of two-particle functions and one-particle densities. The starting point are the atomic orbitals (4.1/3). The usual proceeding is as follows: forming the s.-a. LCAOs according to (5.1/2) and coupling two of these MOs yielding the s.-a. geminals:

$$| [(A\varepsilon e, \varphi_a a)\gamma c, (B\varphi f, \varphi_b b)\delta d]\alpha gr\rangle$$
$$= \{g\}\sqrt{\dim g} \cdot \sum \binom{c^+ d^+ g}{p \; q \; r}^\alpha | (A\varepsilon e, \varphi_a a)\gamma cp\rangle \cdot | (B\varphi f, \varphi_b b)\delta dq\rangle \tag{9.1}$$

For sake of clarity, we repeat this equation omitting the multiplicity indices and readily inserting (5.1) into (9.1):

$$| [(Ae, a)c, (Bf, b)d] gr\rangle$$
$$= \{g\}\sqrt{\dim g} \cdot \sum_{ijmn} \binom{c^+ d^+ g}{p \; q \; r} \cdot K(cp, Aie, am) \cdot K(dq, Bjf, bn) \cdot |Aiam\rangle \cdot |Bjbn\rangle \tag{9.2}$$

Another possible construction of symmetric geminals in the spirit of valence bond theory is as follows: We directly couple the product of the two AOs and size them according to the distance vectors between their atomic centres:

$$| [A, B]Sk, [\varphi_a a, \varphi_b b]\beta ht\rangle$$
$$= \{h\}\sqrt{Z(-ABS)\dim h} \cdot \sum_{ij} \tau\binom{-ABS}{ijk}\binom{a^+ b^+ h}{m \; n \; t}^\beta \cdot |Ai\varphi_a am\rangle \cdot |Bj\varphi_b bn\rangle \tag{9.3}$$

One now notices that these two-centre or edge geminals transform according to the direct product representation $\sigma^S \times h$. The final symmetry-adaption then is achieved by

$$| ([A, B]S\sigma s, [\varphi_a a, \varphi_b b]\beta h)\mu gr\rangle = \sum_{kt} K(\mu gr, Sk\sigma s, ht) | [A, B]Sk, [\varphi_a a, \varphi_b b]\beta ht\rangle \tag{9.4}$$

This construction was already mentioned in [11], eq.(39). But also in the preceeding sections 4 and 8, (9.3/4) is implicitly contained with the difference that not orbitals of two different particles were coupled yielding a s.-a. geminal but the bra and ket orbitals of one particle yielding a s.-a. density. Hence the derivation of (4.4/6) and (8.1) can be achieved in analogy to (9.3/4) by forming consecutively the one-particle densities:

$$[\vec{r} | (Ai\varphi_a a^+, Bj\varphi_b b)\beta ht] = \sqrt{\dim h} \cdot \sum \binom{a \; b^+ h}{m \; n \; t}^\beta \cdot \langle Ai\varphi_a am | \vec{r}\rangle\langle \vec{r} | Bj\varphi_b bn\rangle \tag{9.5}$$

$$[\vec{r} | [A, B]Sk, [\varphi_a a^+, \varphi_b b]\beta ht] = \sqrt{Z(-ABS)}\sum_{ij}\tau\binom{-ABS}{ijk}\cdot[\vec{r}|(Ai\varphi_a a^+, Bj\varphi_b b)\beta ht] \tag{9.6}$$

and finally the s.-a. density:

$$[\vec{r} | ([A, B]S\sigma s, [\varphi_a a^+, \varphi_b b]\beta h)\mu gr] =$$
$$= \sum_{kt} K(\mu gr, Sk\sigma s, ht) \cdot [\vec{r} | [A, B]Sk, [\varphi_a a^+, \varphi_b b]\beta ht] \tag{9.7}$$

The square brackets on the left-hand side are justified by the relation-

ship to the integrals (11.1 and 7). In the case of (4.4/6), subsequently, the density and the tensor operator have to be coupled. Then follows
the integration with respect to $\vec{r}$. The s.-a. density also occurs in the
two-particle integrals.

Both types of geminals (9.1 and 2) must be related by a unitary
transformation, since only the order of different couplings has been
interchanged. The transformation does not depend on the special type
of orbitals, i.e. the quantum numbers φ_a and φ_b, and not on the component r. We then have to calculate the coefficients of the expansion

$$|([A,B]S\sigma s,[\varphi_a a,\varphi_b b]\beta h)\mu g r\rangle$$

$$= \sum_{\varepsilon e \gamma c \varphi f \delta d \alpha} \langle [(A\varepsilon e,a)\gamma c,(B\varphi f,b)\delta d]\alpha g|([A,B]S\sigma s,[a,b]\beta h)\mu g\rangle \qquad (9.8)$$
$$\cdot|[(A\varepsilon e,\varphi_a a)\gamma c,(B\varphi f,\varphi_b b)\delta d]\alpha g r\rangle$$

If we use orthonormalized atomic orbitals like δ-functions the coefficients are the overlap integrals of the geminals (9.1 and 4). Because
of $\langle Ai\varphi_a am|Bk\varphi_b bn\rangle = \delta(A,B)\delta(i,k)\delta(a,b)\delta(m,n)$, we get:

$$\langle [(A\varepsilon e,a)\gamma c,(B\varphi f,b)\delta d]\alpha g|([A,B]S\sigma s,[a,b]\beta h)\mu g\rangle$$
$$= \sqrt{Z(-ABS)}\,\mathrm{dimh}\cdot\mathrm{dimc}\cdot\mathrm{dimd}\cdot\sum_\pi \mathrm{PIs}\begin{pmatrix} -A & B & S \\ \varepsilon_+ & \varphi_+ & \sigma \\ e^+f^+s \end{pmatrix}_\pi \cdot \begin{Bmatrix} c^+d^+g_+ & \alpha \\ e & f & s_+ \\ a & b & h^+ \end{Bmatrix}_\pi \begin{matrix} \alpha \\ \pi \\ \beta \\ \gamma\ \delta\ \mu \end{matrix} \qquad (9.9)$$

The same transformation applies to the densities, and theorem (5.5)
is the result of this recoupling. The geometrical factor (5.7) is equal
to (9.9) except for a different normalization and phase.

10. Pseudo-tetrahedral coefficients

We now treat the pseudo-tetrahedra with numbered vertices just as the oriented triangles in section 7. The principles have been lined out there.

If $\mathfrak{T}$ is the symbol of a set of equivalent pseudo-tetrahedra $\mathfrak{T}_i$, then the eqs.(7.1 to 6) apply with the substitution of $\mathfrak{T}$ for Δ (and $\mathcal{U}$ for θ respectively). We therefore do not repeat them here. We shall term the coefficients of the quadrocentric linear combinations QSALC coefficients, $(\mathfrak{T}_i|\,\alpha ap)$. Quadrocentric linear combinations are s.-a. linear combinations of quadrocentric "objects", especially four-centre integrals. We use the mixtum compositum "quadrocentric" instead of tetracentric, because the letter T has already been used as an abbreviation of tricentric.

The relationship between the tetrahedra and the vertices is again expressed by a topological matrix:

$$\tau\binom{\mathfrak{T}ABCD}{ijklm} = \begin{cases} 1/\sqrt{Z(\mathfrak{T})} & \text{if } \vec{A}_j,\ \vec{B}_k,\ \vec{C}_l,\ \vec{D}_m \text{ in this order are the} \\ & \text{vertices of } \mathfrak{T}_i. \\ 0 & \text{otherwise} \end{cases} \tag{10.1}$$

where $Z(\mathfrak{T})$ is the number of equivalent tetrahedra in the set $\mathfrak{T}$. And again, we associate a vector $\vec{R}_i$ to each pseudo-tetrahedron $\mathfrak{T}_i$ by

$$\vec{R}_i = \sqrt{Z(\mathfrak{T})} \cdot \sum_{jklm} \tau\binom{\mathfrak{T}ABCD}{ijklm} \cdot (\mu_1\vec{A}_j + \mu_2\vec{B}_k + \mu_3\vec{C}_l + \mu_4\vec{D}_m), \tag{10.2}$$

and consequently have

$$(\mathfrak{T}_i|\mathfrak{T}\alpha ap) = (\vec{R}_i|R\alpha ap), \tag{10.3}$$

which is used to calculate the QSALC coefficients. For details cf. section 7.

Again functions of four centres, $F(\vec{A}_j, \vec{B}_k, \vec{C}_l, \vec{D}_m)$, are regarded as functions of the pseudo-tetrahedra,

$$F(\mathfrak{T}_i) = \sqrt{Z(\mathfrak{T})} \cdot \sum_{jklm} \tau\binom{\mathfrak{T}ABCD}{ijklm} \cdot F(\vec{A}_j, \vec{B}_k, \vec{C}_l, \vec{D}_m), \tag{10.4}$$

and can be expanded in QSALC coefficients:

$$F(\mathfrak{T}_i) = (\mathfrak{T}_i|F) = \sum_{\alpha a}(\mathfrak{T}_i|\mathfrak{T}\alpha ap)(\mathfrak{T}\alpha ap|F) \tag{10.5}$$

with

$$(\mathfrak{T}\alpha ap|F) = \sum_i(\mathfrak{T}\alpha ap|\mathfrak{T}_i)(\mathfrak{T}_i|F). \tag{10.6}$$

The topological matrices (10.1) have the orthogonality relations:

$$\sum_{jklm} \tau\binom{\mathfrak{T}ABCD}{ijklm}\tau\binom{\mathcal{U}ABCD}{njklm} = \delta(\mathfrak{T},\mathcal{U})\delta(i,n)/Z(\mathfrak{T}) \tag{10.7}$$

$$\sum_{\mathfrak{T}i} Z(\mathfrak{T})\tau\binom{\mathfrak{T}ABCD}{ijklm}\tau\binom{\mathfrak{T}ABCD}{irstu} = \delta(j,r)\delta(k,s)\delta(l,t)\delta(m,u) \tag{10.8}$$

Again we can reduce the topological matrix (10.1) by a factorization. For this purpose, we characterize the pseudo-tetrahedra by two opposite edge vectors, which link the vertices 1 and 3, and 2 and 4 respectively. This relationship is expressed by:

$$\tau^3\binom{\mathcal{S}ST}{ikl} = \begin{cases} 1/\sqrt{Z(\mathcal{S})} & \text{if } \vec{S}_k \text{ links the vertices 1 and 2, and } \vec{T}_1 \text{ links} \\ & \text{the vertices 2 and 4.} \\ 0 & \text{otherwise} \end{cases} \qquad (10.9)$$

The factorization then is given by:

$$\tau\binom{\mathcal{S}ABCD}{ijklm} = \sum_{Ss}\sum_{Tt}\tau^3\binom{\mathcal{S}ST}{st}\sqrt{Z(-ACS)}\tau\binom{-ACS}{jls}\sqrt{Z(-BDT)}\tau\binom{-BDT}{kmt} \qquad (10.10)$$

The topological matrix τ^3 now is treated just as τ^2 in section 7. The orthogonality relations are:

$$\sum_{st}\tau^3\binom{\mathcal{S}ST}{ist}\tau^3\binom{\mathcal{U}ST}{jst} = \delta(\mathcal{S},\mathcal{U})\delta(i,j)/Z(\mathcal{S}) \qquad (10.11)$$

$$\sum_{\mathcal{S}i}Z(\mathcal{S})\tau^3\binom{\mathcal{S}ST}{ist}\tau^3\binom{\mathcal{S}ST}{ipq} = \delta(s,p)\delta(t,q) \qquad (10.12)$$

The transformation into the s.-a. basis yields:

$$\tau^3\begin{pmatrix}\mathcal{S}ST\\ \alpha\sigma\tau \\ abc \\ xyz\end{pmatrix} = \sum_{ist}\tau^3\binom{\mathcal{S}ST}{ist}(\mathcal{S}_1|\mathcal{S}\alpha ax)(\vec{S}_s|S\sigma by)(\vec{T}_t|T\tau cz) \qquad (10.13)$$

with the subsequent factorization

$$\tau^3\begin{pmatrix}\mathcal{S}ST\\ \alpha\sigma\tau \\ abc \\ xyz\end{pmatrix} = \sum_\varepsilon PIs^3\begin{vmatrix}\mathcal{S}ST\\ \alpha\sigma\tau \\ abc\end{vmatrix}_\varepsilon \cdot \binom{abc}{xyz}\varepsilon, \qquad (10.14)$$

where the "polyhedral isoscalar of the third kind" is given by:

$$PIs^3\begin{vmatrix}\mathcal{S}ST\\ \alpha\sigma\tau \\ abc\end{vmatrix}_\varepsilon = \sum_{ist}\tau^3\binom{\mathcal{S}ST}{ist}(\mathcal{S}_1|\mathcal{S}\alpha ax)(\vec{S}_s|S\sigma by)(\vec{T}_t|T\tau cz)\binom{abc}{xyz}\varepsilon^* \qquad (10.15)$$

From (10.11/12), one derives the orthogonality relations:

$$\sum_{\sigma b}\sum_{\tau c\varepsilon}PIs^3\begin{vmatrix}\mathcal{S}ST\\ \alpha\sigma\tau \\ abc\end{vmatrix}_\varepsilon^* PIs^3\begin{vmatrix}\mathcal{U}ST\\ \beta\sigma\tau \\ abc\end{vmatrix}_\varepsilon = \delta(\mathcal{S},\mathcal{U})\delta(\alpha,\beta)dima/Z(\mathcal{S}) \qquad (10.16)$$

$$\sum_{\mathcal{S}\alpha}Z(\mathcal{S})PIs^3\begin{vmatrix}\mathcal{S}ST\\ \alpha\sigma\tau \\ abc\end{vmatrix}_\varepsilon^* PIs^3\begin{vmatrix}\mathcal{S}ST\\ \alpha\sigma\tau \\ abc\end{vmatrix}_\eta = \delta(\sigma,\sigma)\delta(b,b)\delta(\tau,\tau)\delta(c,c)\delta(\varepsilon,\eta)\cdot dima \qquad (10.17)$$

Without the factorization (10.10) a rather complex polyhedral isoscalar results from (10.1):

$$PIs\begin{vmatrix}\mathcal{S}ABCD\\ \tau\alpha\beta\gamma\delta \\ tabcd\end{vmatrix}_{\varepsilon e\varphi f\mu} = \sum_{qi jkl}\tau\binom{\mathcal{S}ABCD}{qijkl}(\mathcal{S}_q|\mathcal{S}\tau tp_t)(\vec{A}_i|A\alpha ap_a)(\vec{B}_j|B\beta bp_b)$$
$$\cdot(\vec{C}_k|C\gamma cp_c)(\vec{D}_1|D\delta dp_d)dime\cdot dimf\binom{a\ c\ e}{p_a p_c p_e}\varepsilon^*\binom{f\ b\ d}{p_f p_b p_d}\varphi^*\binom{t\ e\ f}{p_t p_e p_f}\mu^* \qquad (10.18)$$

The subsequent application of (1o.10) splits up this coefficient as follows:

$$PIs\begin{vmatrix}\mathcal{S}ABCD\\ \tau\alpha\beta\gamma\delta \\ tabcd\end{vmatrix}_{\varepsilon e\varphi f\mu} = \sum_{S\varepsilon T\varphi}\sqrt{Z(-ACS)Z(-BDT)}\cdot PIs\begin{vmatrix}-ACS\\ \alpha\gamma e \\ ace\end{vmatrix}_\varepsilon PIs\begin{vmatrix}-BDT\\ \beta\delta\varphi \\ bdf\end{vmatrix}_\varphi PIs^3\begin{vmatrix}\mathcal{S}ST\\ \tau e\varphi \\ tef\end{vmatrix}_\mu \qquad (10.19)$$

11. Two-particle interaction

11.1. The four-centre integrals

The interaction of the molecular electrons is, except for relativistic effects, represented by the scalar operator $1/r_{12}$, which in general yields four-centre integrals over the atomic orbitals. As now can be foreseen, these integrals can be expanded in QSALC coefficients according to (10.5).

With respect to the four orbitals there are different coupling modes. Either the orbitals in the bra and in the ket are coupled separately yielding two-centre geminals, or the orbitals referring to the same particle are coupled resulting in two s.-a. densities. Whereas the first coupling mode takes into account the separation of the interaction according to electron pairs, the second mode is more capable for the calculation of the integrals. This aspect, being more important in the present context, is accentuated by passing over to the notation usual in quantum chemistry (cf. [35, 36]):

$$\langle Ai\varphi_a am, Bj\varphi_b bn | r_{12}^{-1} | Ck\varphi_c cp, Dl\varphi_d dq \rangle = [Ai\varphi_a a^+m, Ck\varphi_c cp | r_{12}^{-1} | Bj\varphi_b b^+n, Dl\varphi_d dq] \tag{11.1}$$

Since the square brackets are defined without complex conjugation, the conjugated representations a^+ and b^+ show up explicitly.

Using the two-centre densities (9.5), the integrals (11.1) are reduced to

$$[Ai\varphi_a a^+m, Ck\varphi_c cp | r_{12}^{-1} | Bj\varphi_b b^+n, Dl\varphi_d dq] \tag{11.2}$$

$$= \sum_{\epsilon e \varphi f} [Ai\varphi_a a^+_a, Ck\varphi_c c) \epsilon er | r_{12}^{-1} | (Bj\varphi_b b^+_b, Dl\varphi_d d) \varphi fs] \sqrt{\dim e \cdot \dim f} \begin{pmatrix} a^+ & c & e \\ m & p & r \end{pmatrix} \epsilon \begin{pmatrix} f^+ & b^+ & d \\ s & n & q \end{pmatrix} \varphi$$

Note that, in contrast to the one-centre integrals of scalar operators, the representations e and f may be different.

The integrals of the right-hand side now are further reduced by the factorization theorem:

$$[(Ai\varphi_a a^+, Ck\varphi_c c) \epsilon er | r_{12}^{-1} | (Bj\varphi_b b^+, Dl\varphi_d d) \varphi fs] \tag{11.3}$$

$$= \sum_{\mathcal{F} \tau \mu g} \sum \tau \begin{pmatrix} \mathcal{F} ABCD \\ tijkl \end{pmatrix} \sqrt{Z(\mathcal{F})} [(A\varphi_a a^+, B\varphi_b b) \epsilon e \| r_{12}^{-1} \| (B\varphi_b b^+, D\varphi_d d) \varphi f]^\mu_{\mathcal{F}\tau g} \begin{pmatrix} e^+ f & g^+ \\ r & s & u \end{pmatrix} \mu (\mathcal{F}_t | \mathcal{F}_\tau gu)$$

By this relation, the quadrocentric, reduced matrix elements, QRM, are defined. They give the following informations: A QRM belongs to a set $\mathcal{F}$ of tetrahedra, the representation of which is reduced to g (multiplicity τ) by symmetry-adaption; the two-centre densities are coupled yielding the representations e (multiplicity ϵ) and f (multiplicity φ); finally the coupling of e and f has to yield representation g again (with multiplicity μ).

The proof is as for (4.4/6) or (8.1). Omitting the multiplicities again accentuates the essential structure:

$$[Aia^+m, Ckcp | r_{12}^{-1} | Bjb^+n, Dldq] \tag{11.4}$$

$$= \sum_{ef} [(Aia^+, Ckc)er | r_{12}^{-1} | (Bjb^+, Dld)fs] \sqrt{\text{dime} \cdot \text{dimf}} \begin{pmatrix} a^+ c & e \\ m & p & r \end{pmatrix} \begin{pmatrix} f^+ b^+ d \\ s & n & q \end{pmatrix}$$

and

$$[(Aia^+, Ckc)er | r_{12}^{-1} | (Bjb^+, Dld)fs] \tag{11.5}$$

$$= \sum_{\Im tg} \tau \begin{pmatrix} \Im ABCD \\ tijkl \end{pmatrix} \sqrt{Z(\Im)} [(Aa^+, Cc)e \| r_{12}^{-1} \| (Bb^+, Dd)f]_{\Im g} \begin{pmatrix} e^+ f & g \\ r & s & u \end{pmatrix} (\Im_t | \Im gu)$$

The sums for $\Im$ and t contain one non-zero summand only, since they sort out the right pseudo-tetrahedron. Only that for g (and u) is a true sum.

11.2. The matrix elements of the s.-a. molecular orbitals and geminals

The next aim is the calculation of the matrix elements of the operator $1/r_{12}$ with respect to the s.-a. LCAO-MOs (5.1) and the geminals (9.1).

Since now four centres are involved, we need a more economical and compact notation to manage all the quantum numbers. We rewrite (5.1) as follows:

$$|\Lambda \dot{a} p_a\rangle = |(A \dot{a} \dot{a}, \alpha a) \dot{a} \dot{a} p_a\rangle = \sum_{lm_a} K(\dot{a} \dot{a} p_a, Ai \dot{a} \dot{a}, am_a) \cdot |Ai\alpha am_a\rangle \tag{11.6}$$

The letters a and α always point to the centres of set A (b and β to B respectively). Λ is a collective index for $(A \dot{a} \dot{a}, \alpha a) \dot{a}$. We further introduce $\Lambda^+ = (A \dot{a} \dot{a}^+, \alpha a^+) \dot{a}$.

As for the integrals (11.1), there are two possible notations of the integrals of the MOs (11.6):

$$\langle \Lambda \dot{a} p_a, \Bslash \dot{b} p_b | r_{12}^{-1} | \Cslash \dot{c} p_c, \Dslash d p_d \rangle = [\Lambda^+ \dot{a}^+ p_a, \Cslash \dot{c} p_c | r_{12}^{-1} | \Bslash^+ \dot{b}^+ p_b, \Dslash d p_d] \tag{11.7}$$

Consequently, there are two types of reduced matrix elements. In the particle-coupling, the application of the WET yields:

$$\langle \Lambda \dot{a} p_a, \Bslash \dot{b} p_b | r_{12}^{-1} | \Cslash \dot{c} p_c, \Dslash d p_d \rangle \tag{11.8}$$

$$= \sum_{\sigma \mu s} \langle (\Lambda \dot{a}, \Bslash \dot{b}) \sigma s \| r_{12}^{-1} \| (\Cslash \dot{c}, \Dslash d) \mu s \rangle \cdot \text{dims} \begin{pmatrix} \dot{a}^+ \dot{b}^+ s \\ p_a p_b p_s \end{pmatrix} \sigma \begin{pmatrix} s^+ \dot{c} & d \\ p_s p_c p_d \end{pmatrix} \mu$$

and in the density coupling:

$$\langle \Lambda \dot{a} p_a, \Bslash \dot{b} p_b | r_{12}^{-1} | \Cslash \dot{c} p_c, \Dslash d p_d \rangle \tag{11.9}$$

$$= \sum_{\tau \pi t} [(\Lambda^+ \dot{a}^+, \Cslash \dot{c}) \tau t \| r_{12}^{-1} \| (\Bslash^+ \dot{b}^+, \Dslash d) \pi t] \cdot \text{dimt} \begin{pmatrix} \dot{a}^+ \dot{c} & t \\ p_a p_c p_t \end{pmatrix} \tau \begin{pmatrix} t^+ \dot{b}^+ d \\ p_a p_c p_t \end{pmatrix} \pi$$

Both types of reduced matrix elements are interrelated by 6j symbols (cf. [17]):

$$\langle (\Lambda \dot{a}, \Bslash \dot{b}) \sigma s \| r_{12}^{-1} \| (\Cslash \dot{c}, \Dslash d) \mu s \rangle \tag{11.10}$$

$$= \sum_{\tau \pi t} \begin{Bmatrix} \dot{a} & \dot{b} & s^+ \\ \dot{d} & \dot{c} & t \end{Bmatrix}_{\tau \pi \mu \sigma} \{\dot{c} d s^+ \mu\} \{\dot{b}^+ d t \pi\} \text{dimt} \cdot [(\Lambda^+ \dot{a}^+, \Cslash \dot{c}) \tau t \| r_{12}^{-1} \| (\Bslash^+ \dot{b}^+, \Dslash d) \pi t]$$

The reduced matrix elements in the particle coupling are also immediately related to the integrals of the s.-a. geminals (9.1) now taking the form:

$$|(\hat{A}\dot{a},\not{B}\dot{b})\sigma sp_s\rangle = \{s\}\sqrt{\text{dims}}\cdot\sum(\overset{\dot{a}+\dot{b}+s}{p_ap_bp_s})^\sigma\cdot|\hat{A}\dot{a}p_a\rangle\cdot|\not{B}\dot{b}p_b\rangle \qquad (11.11)$$

The interaction integrals of these geminals are:

$$\langle(\hat{A}\dot{a},\not{B}\dot{b})\sigma sp_s|r_{12}^{-1}|(\not{C}\dot{c},\not{D}\dot{d})\mu sp_s\rangle$$
$$= \delta(s,s)\delta(p_s,p_s)\text{dims}^{-1}\langle(\hat{A}\dot{a},\not{B}\dot{b})\sigma s\|r_{12}^{-1}\|(\not{C}\dot{c},\not{D}\dot{d})\mu s\rangle, \qquad (11.12)$$

where the reduced matrix elements on the right are those of eq.(11.6).

The reduced matrix elements defined in (11.8/9) now must be traced back to the QRMs. Since (11.9) as well as (11.3) are based on the density coupling, their interrelation is more direct:

$$[(\hat{A}^+\dot{a}^+,\not{C}\dot{c})\tau t\|r_{12}^{-1}\|(\not{B}^+\dot{b}^+,\not{D}\dot{d})\pi t]$$
$$= \sum_{\varepsilon\varepsilon\varphi I}\sum_{\mu\sigma g\mathfrak{I}}GEO_3(\varepsilon\varepsilon\varphi I\mu,\mathfrak{I}\sigma g)\cdot[(A\alpha a^+,C\gamma c)\varepsilon e\|r_{12}^{-1}\|(B\beta b^+,D\delta d)\varphi f]^\mu_{\mathfrak{I}\sigma g} \qquad (11.13)$$

The geometrical factor is derived as follows. We invert (11.9):

$$[(\hat{A}^+\dot{a}^+,\not{C}\dot{c})\tau t\|r_{12}^{-1}\|(\not{B}^+\dot{b}^+,\not{D}\dot{d})\pi t]$$
$$= \sum(\overset{\dot{a}+\dot{c}\ t}{p_ap_cp_t})\tau^*(\overset{t+\dot{b}+\dot{d}}{p_tp_bp_d})\pi^*\langle\hat{A}\dot{a}p_a,\not{B}\dot{b}p_b|r_{12}^{-1}|\not{C}\dot{c}p_c,\not{D}\dot{d}p_d\rangle, \qquad (11.14)$$

where the matrix elements of the right-hand side, because of (11.6), are given by:

$$\langle\hat{A}\dot{a}p_a\cdot\not{B}\dot{b}p_b|r_{12}^{-1}|\not{C}\dot{c}p_c,\not{D}\dot{d}p_d\rangle = \sum_{ijklm_x}\sum K(\dot{a}\dot{a}p_a,A i\dot{a}\dot{a}',am_a)^*\cdot K(\dot{\beta}\dot{b}p_b,Bj\beta\dot{b}',bm_b)^*$$
$$\cdot K(\dot{\gamma}\dot{c}p_c,Ck\gamma\dot{c}',cm_c)\cdot K(\dot{\delta}\dot{d}p_d,Dl\delta\dot{d}',dm_d)\langle Ai\alpha am_a,Bj\beta bm_b|r_{12}^{-1}|Ck\gamma cm_c,Dl\delta dm_d\rangle \qquad (11.15)$$

We now insert (11.1-3) into (11.15) and the result into (11.14). These substitutions yield the factor:

$$GEO_3(\varepsilon\varepsilon\varphi I\mu,\mathfrak{I}\sigma g) = \sum_{ijklm_xq}\sum(\overset{\dot{a}+\dot{c}\cdot t}{p_ap_cp_t})\tau^*\cdot(\overset{t+\dot{b}+\dot{d}}{p_tp_bp_d})\pi^*\cdot K(\alpha ap_a,Ai\dot{a}\dot{a}',am_a)^*$$
$$\cdot K(\beta bp_b,Bj\beta\dot{b}',bm_b)^*\cdot K(\gamma cp_c,Ck\gamma\dot{c}',cm_c)\cdot K(\delta dp_d,Dl\delta\dot{d}',dm_d)$$
$$\cdot\sqrt{\text{dime}\cdot\text{dimf}}\cdot(\overset{\dot{a}+\dot{c}\ e}{m_am_cr})\varepsilon(\overset{\dot{b}+\dot{d}\ f}{m_bm_ds})\varphi(\overset{e+f+g}{r\ s\ u})^\mu\sqrt{Z(\mathfrak{I})}\tau(\overset{\mathfrak{I}ABCD}{qijkl})\cdot(\mathfrak{I}_q|\mathfrak{I}\sigma gu) \qquad (11.16)$$

This formula contains nine 3jm symbols, which now are rearranged by a two-fold application of the rule of de Shalit (2.54) and the subsequent application of (2.41):

$$GEO_3(\varepsilon\varepsilon\varphi I\mu,\mathfrak{I}\sigma g) =$$
$$= (Z(\mathfrak{I})\text{dim}\hat{a}\cdot\text{dim}\hat{b}\cdot\text{dim}\hat{c}\cdot\text{dim}\hat{d}\cdot\text{dime}\cdot\text{dimf})^{1/2}\sum_{ijklqn\varepsilon e\eta\varphi I\mu'}\sum\sum\text{dime}'\cdot\text{dimf}'$$
$$\cdot\begin{Bmatrix}a'&d+e'\\a\ &c+e\\\dot{a}+\dot{c}&t\end{Bmatrix}\overset{\varepsilon'}{\underset{\varepsilon}{\tau}}\begin{Bmatrix}f'&b'&d+\\f+b&d+\\t+\dot{b}+\dot{a}\end{Bmatrix}\overset{\varphi'}{\underset{\pi}{\varphi}}\begin{Bmatrix}e\ f\ g\\f+e'\ t+\end{Bmatrix}\eta\eta\mu\mu\cdot\{\dot{e}e+t\eta\}(\overset{a'\ d+e'}{r_ar_ce})\varepsilon'(\overset{f'\ b'\ d+}{r_fr_bd})\varphi'(\overset{f'\ e'\ g}{r_fr_eu})\mu'^* \qquad (11.17)$$
$$\cdot(\overrightarrow{A_i}|A\dot{a}\dot{a}'r_a)\cdot(\overrightarrow{B_j}|B\beta\dot{b}'r_b)\cdot(\overrightarrow{C_k}|C\gamma\dot{c}'r_c)\cdot(\overrightarrow{D_l}|D\delta\dot{d}'r_d)\cdot\tau(\overset{\mathfrak{I}ABCD}{qijkl})\cdot(\mathfrak{I}_q|\mathfrak{I}\sigma gu)$$

The last nine factors form a scalar being nearly identical with the polyhedral isoscalar (10.18):

$$GEO_3(\varepsilon e\varphi f\mu, \mathfrak{F}\sigma g) = (Z(\mathfrak{F})dim\dot{a}\cdot dim\dot{b}\cdot dim\dot{c}\cdot dimd\cdot dime\cdot dimf)^{1/2}\sum_{\eta\bar{\varepsilon}\bar{\varepsilon}\eta\varphi f\mu}\sum \{\dot{e}e^+t\eta\}$$

(11.18)

$$\cdot\{\dot{e}\dot{f}g\mu'\}\cdot\begin{Bmatrix} a' & \dot{c}^+\dot{e}' \\ a_+ & c_+^+e^+ \\ \dot{a} & \dot{c} & t \\ \alpha & \gamma & \eta \end{Bmatrix}\begin{matrix}\varepsilon' \\ \varepsilon \\ \tau\end{matrix}\cdot\begin{Bmatrix} f' & b' & d'^+ \\ f_+^+ & b_+ & d_+^+ \\ t_+ & \dot{b} & \dot{d} \\ \eta' & \beta & \delta \end{Bmatrix}\begin{matrix}\varphi' \\ \varphi \\ \pi\end{matrix}\cdot\begin{Bmatrix} e_+ & f & g \\ f_+^+ & \dot{e} & t^+ \end{Bmatrix}_{\eta\eta\mu\mu}\cdot PIs\begin{vmatrix} \mathfrak{F} & A & B & C & D \\ \sigma & \alpha' & \beta' & \gamma' & \delta' \\ g & a^+b^+c' & d' \end{vmatrix}_{\dot{\varepsilon}e\varphi f\mu'}$$

As a result of the discussion in section 9, we expect that this geometrical factor contains twice a recoupling of the densities according
to (9.9) or (5.7). This shows up, if we split up the polyhedral isoscalar by (10.19):

$$GEO_3(\varepsilon e\varphi f\mu, \mathfrak{F}\sigma g) = F\cdot\sum_{\mathfrak{F}\mathcal{A}T\nu\bar{\eta}\bar{\varepsilon}\bar{\varepsilon}\eta\varphi f\mu'}\sum\sum\{\dot{e}e^+t\eta\}\{\dot{e}\dot{f}g\mu\}\cdot\begin{Bmatrix} a' & \dot{c}^+\dot{e}' \\ a_+ & c_+^+e^+ \\ \dot{a} & \dot{c} & t \\ \alpha & \gamma & \eta \end{Bmatrix}\begin{matrix}\varepsilon' \\ \varepsilon \\ \tau\end{matrix}\cdot\begin{Bmatrix} f' & b' & d'^+ \\ f_+^+ & b_+ & d \\ t_+ & \dot{b} & \dot{d} \\ \eta' & \beta & \delta \end{Bmatrix}\begin{matrix}\varphi' \\ \varphi \\ \pi\end{matrix}$$

(11.19)

$$\cdot\begin{Bmatrix} e_+ & f & g \\ f_+^+ & \dot{e} & t^+ \end{Bmatrix}_{\eta\eta\mu\mu}\cdot PIs\begin{vmatrix} -A & C & S \\ \alpha' & \gamma' & \mathcal{A} \\ a^+c' & \dot{e}^+ \end{vmatrix}_{\varepsilon'}\cdot PIs\begin{vmatrix} -B & D & T \\ \beta' & \delta' & \mathcal{J} \\ b^+d' & f^+ \end{vmatrix}_{\varphi'}\cdot PIs^3\begin{vmatrix} \mathfrak{F}ST \\ \sigma\mathcal{A}\nu \\ gef \end{vmatrix}_{\mu'}$$

with the abbreviation $F = \sqrt{Z(\mathfrak{F})Z(-ACS)Z(-BDT)dim\dot{a}\cdot dim\dot{b}\cdot dim\dot{c}\cdot dimd\cdot dime\cdot dimf}$

The comparison with (5.7) then yields:

$$GEO_3(\varepsilon e\varphi f\mu, \mathfrak{F}\sigma g) = \{\dot{c}\}\{\dot{d}\}\sqrt{Z(\mathfrak{F})}dime\cdot dimf\cdot\sum_{\mathfrak{F}\mathcal{A}T\nu\bar{\eta}\bar{\varepsilon}\bar{\eta}\varphi f\mu'}\sum\{\dot{e}e^+t\eta\}\{\dot{e}\dot{f}g\mu\}\{\dot{a}^+\dot{c}t\tau\}$$

(11.20)

$$\cdot\{\dot{b}^+dt^+\pi\}\cdot\begin{Bmatrix} e_+ & f & g \\ f_+^+ & \dot{e} & t^+ \end{Bmatrix}_{\eta\eta\mu\mu}\cdot PIs^3\begin{vmatrix} \mathfrak{F}ST \\ \sigma\mathcal{A}\nu \\ gef \end{vmatrix}_{\mu}\cdot GEO_1(x_1,y_1,z_1)\cdot GEO_1(x_2,y_2,z_2)$$

with the compound arguments: $x_1=(AaCct)$, $y_1=(\alpha'\dot{a}\dot{a}\dot{a}, \gamma'\dot{c}\gamma\dot{c}, \tau)$, $z_1=(\varepsilon e\eta S\mathcal{A}e^+)$,
$x_2=(BbDdt^+)$, $y_2=(\beta'\dot{b}\beta\dot{b}, \delta'\dot{d}\delta\dot{d}, \pi)$, and $z_2=(f\varphi\eta T\nu f^+)$.

Finally we sum up the results omitting again the multiplicity indices. Because the many-particle matrix elements of the two-particle interaction can be traced back to the matrix elements of the s.-a. geminals using the coefficients of fractional parentage [27, 17], we start
with the matrix elements of the geminals. According to (11.12/10/13)
we have:

$$\langle(A\dot{a},B\dot{b})s\|r_{12}^{-1}\|(C\dot{c},D\dot{d})s\rangle$$

(11.21)

$$=\sum_{\text{tef}\mathfrak{F}g}\sum\begin{Bmatrix} \dot{a} & \dot{b} & s^+ \\ \dot{d} & \dot{c} & t \end{Bmatrix}\{\dot{c}ds^+\}\{\dot{b}dt\}dimt\cdot GEO_3(ef,\mathfrak{F}g)\cdot\left[(Aa^+,Cc)e\|r_{12}^{-1}\|(Bb^+,Dd)f\right]_{\mathfrak{F}g}$$

with

$$GEO_3(ef,\mathfrak{F}g) = \{\dot{c}\}\{\dot{d}\}\sqrt{Z(\mathfrak{F})}dime\cdot dimf\cdot\sum_{STef}\{\dot{e}e^+t\}\{\dot{e}\dot{f}g\}\{\dot{a}^+\dot{c}t\}\{\dot{b}^+dt^+\}\begin{Bmatrix} e_+ & f & g \\ f_+^+ & \dot{e} & t^+ \end{Bmatrix}$$

(11.22)

$$\cdot PIs^3(\begin{smallmatrix}\mathfrak{F}ST \\ gef\end{smallmatrix})\cdot GEO_1((AaCct),(\dot{a}\dot{a},\dot{c}\dot{c}),(eSe^+))\cdot GEO_1((BbDdt^+),(\dot{b}\dot{b},\dot{d}\dot{d}),(fTf^+))$$

Later on from a more general point of view it will turn out, that
the factor GEO_3 is even a triple combination of the factors of type
GEO_1.

12. Complete bases of irreducible representations

Having delineated the principal structures proper to the symmetric polyhedra, we will implement the theory in the following sections by details concerning the determination of the newly defined coefficients, or of the invariants BRM, TRM, and QRM from specified approximations or ab initio formulae. The starting point is the systematic reconstruction of capable, s.-a. basis sets with respect to the centre of symmetry.

12.1. Generalization of Kopsky's theorem

All s.-a. functions of the translation group can be factorized as $\varphi_{nk}(x) = u_{nk}(x) \cdot \exp(ikx)$ [37] and those of the rotation group as $\varphi_{nlm}(\vec{r})$ $= R_{nl}(r) \cdot \langle\vec{r}|lm\rangle$, where $u_{nk}(x)$ and $R_{nl}(r)$ are scalar functions of the respective group. This means that there is one function $\exp(ikx)$ or one set of functions $\langle\vec{r}|lm\rangle$, which, combined with an infinite set of scalar functions, constitutes the complete set of basis functions.

This suggests the question, wether there is an analogue in the case of point groups. Of course, the s.-a. spherical harmonics (3.12) constitute the complete basis $R_{nl}\langle\vec{r}|l\gamma cp\rangle$, and most books on group theory are content with it. But this basis is not the wanted one; since for each representation c there are several, moreover infinitely many values of l and γ. Thus there are required infinitely many sets of functions instead of one. The perfect analogue would be one set of standard functions $\langle\vec{r}|st.ap\rangle$ so that $R_{na}(\vec{r})\langle\vec{r}|st.ap\rangle$ constitutes the complete set. It turns out that this perfect analogy exists only for the one-dimensional, irreducible representations, whereas for the many-dimensional, irreducible representations several, standardized sets $\langle\vec{r}|st.\alpha ap\rangle$ are required, the munber of which is equal to dima, i.e. the multiplicity index α runs from 1 to dima.

Whereas the construction of s.-a. spherical harmonics or equivalently of. s.-a. homogeneous polynoms in x, y, z is worked out in long lists of functions [38] and is included in most books on group theory, the present problem has obviously not attracted much attention. Nevertheless it turns out that every arbitrary, s.-a. function $\langle\vec{r}|\varphi ap\rangle$ of species a can be expanded in a finite number of standard functions:

$$\langle\vec{r}|\varphi ap\rangle = \sum_{\alpha=1}^{\text{dima}} R_{\alpha a}^{\varphi}(\vec{r}) \cdot \langle\vec{r}|st.\alpha ap\rangle \qquad (12.1)$$

An exception from this indifference is a late paper of Kopsky [39]. We learn from it that the problem previously has been passed casually [40, 41], although - as Kopsky shows - partly in a misleading manner.

Since Kopsky's theorem 1, the only one used here, is equivalent to eq. (12.1), we take this for granted. But Kopsky and his predecessors

confined themselves to the existence of the linearly independent functions and the expansion (12.1). They did not show, how the scalar func-
tions $R_{\alpha a}^{\varphi}(\vec{r})$ are determined by a given, s.-a. function $\langle\vec{r}|\varphi ap\rangle$. On the
contrary, we can provide a system of standardized functions with stron-
ger properties allowing the inversion of (12.1) and a systematic calculation of the SALC (including TSALC and QSALC) coefficients.

We can regard (12.1) as a special case of the expansion (3.16). Being unaware about Kopsky's theorem we could prove (12.1) in this way.
Every position vector $\vec{r}$ together with its rotated images $g\vec{r}$ makes up
an equivalent set $R=\{g\vec{r}$ with $g\epsilon G\}$ in the sense of section 3. This means
that the standard functions are proportional to the SALC coefficients
for general positions (i.e. positions being invariant to no symmetry
operation). The normalization is chosen different allowing for continu-
ity if $\vec{r}$ approaches an element of symmetry. In analogy to (3.17), we
use the scalar product

$$(F|G) = \sum_{g\epsilon G}\langle F|g\vec{r}\rangle\langle g\vec{r}|G\rangle \tag{12.2}$$

with respect to the discrete set R in order to orthonormalize the standard functions.

We now compile the properties of the standard functions in the following theorem: For each irreducible representation a of a point group
G there is a set of standardized functions $\langle\vec{r}|st.\alpha ap\rangle$ having the pro-
perties:

1) $\alpha = 1, 2,\ldots,$ dima.

2) $\sum_{g\epsilon G}\langle st.\alpha ap|g\vec{r}\rangle\langle g\vec{r}|st.\beta bq\rangle = \delta(\alpha,\beta)\delta(a,b)\delta(p,q)\cdot\mu(\alpha a,\vec{r})$ (12.3)

   with $\mu(\alpha a,g\vec{r}) = \mu(\alpha a,\vec{r})$ (12.4)

   and $\mu(\alpha a,\vec{r})\geqslant 0$, where $\mu(\alpha a,\vec{r}) \neq 0$ if $\vec{r}$ in general position (12.5)

3) $\sum_{\alpha ap}\langle g\vec{r}|st.\alpha ap\rangle\mu(\alpha a,\vec{r})^{-1}\langle st.\alpha ap|h\vec{r}\rangle = \delta(g,h)$ (12.6)

   for every $\vec{r}$ in general position.

4) Every function $\langle\vec{r}|\varphi ap\rangle$ of species a (component p) is representable
   by the expansion

$$\langle\vec{r}|\varphi ap\rangle = \sum_{\alpha=1}^{\text{dima}} R_{\alpha a}^{\varphi}(\vec{r})\langle\vec{r}|st.\alpha ap\rangle , \tag{12.7}$$

   where the scalar functions $R_{\alpha a}^{\varphi}(\vec{r})$ are defined by:

$$\mu(\alpha a,\vec{r})\cdot R_{\alpha a}^{\varphi}(\vec{r}) = \sum_{g\epsilon G}\langle st.\alpha ap|g\vec{r}\rangle\langle g\vec{r}|\varphi ap\rangle \tag{12.8}$$

5) Among other possibilities, the standard functions can be chosen ho-
   mogeneous in x, y, z. But in general, the scalar functions are no
   polynomes.

The proof is based on Kopsky's theorem 1. There are precisely dima li-
nearly independent functions, which we may term $\langle\vec{r}|nap\rangle$ (with n=1,...,
dima). With respect to the set $R=\{g\vec{r}$ with $g\epsilon G\}$, we can express the linear independence by Gram's determinant [42]:

$$\det\left|(\text{map}|\text{nap})\right| > 0, \text{if } \vec{r} \text{ in general position,}$$

with
$$(\text{map}|\text{nap}) = \sum_{g \in G} \langle\text{map}|g\vec{r}\rangle\langle g\vec{r}|\text{nap}\rangle. \tag{12.9}$$

If we admit points of symmetry like $\vec{r}=\vec{S}_i$, the rank of the determinant must decrease because of $n(S,a) < \text{dima}$. Thus in general, we have only
$$\det\left|(\text{map}|\text{nap})\right| \geqslant 0. \tag{12.10}$$

We now define the standard functions by the diagonalization of the matrix $X_{mn}(\vec{r})=(\text{map}|\text{nap})$ by a non-singular transformation:

$$\langle\vec{r}|\text{st.}\alpha\text{ap}\rangle = \sum_{n=1}^{\text{dima}} u_{\alpha n}(\vec{r})\langle\vec{r}|\text{nap}\rangle \tag{12.11}$$

This yields our statements (12.3 to 5). The zeroes of $\det\left|X_{mn}(\vec{r})\right|$ and of its eigenvalues $\mu(\alpha a,\vec{r})$ prevent us from normalizing the functions without destroying the continuity in the whole range of $\vec{r}$. Because of Kopsky's theorem 1 the functions $\langle\vec{r}|\text{nap}\rangle$ and therefore the functions $\langle\vec{r}|\text{st.}\alpha\text{ap}\rangle$, too, are complete. The completeness relation is (12.6), from which follows (12.7). Inverting this by (12.3) yields (12.8).

It is convenient, but not necessary, to choose the standard functions as homogeneous polynomes. According to Kopsky's theorem 2, the number of independent polynomes is equal to or higher than dima. Hence the number of polynomes is sufficient. We can take them from the lists given in [38]. A general method for the generation of polynomes is described in [43]. Furthermore care has to be taken that the transformation (12.11) preserves the polynomial property. This is possible for instance using Schmidt's orthogonalization without normalization, i.e. without any divisions. Finally the homogeneous polynomes $\langle\vec{r}|\text{st.}\alpha\text{ap}\rangle$ and the scalar, homogeneous polynomes $\mu(\alpha a,\vec{r})$ are fixed only up to a further transformation of type (12.11).

12.2. Applications

We now come to the applications of the theorem. The relation (12.7), of course, applies to the s.-a. spherical harmonics (3.19):

$$\langle\vec{r}|1\alpha\text{ap}\rangle = \sum_{\gamma=1}^{\text{dima}} R_{\gamma a}^{1\alpha}(\vec{r})\langle\vec{r}|\text{st.}\gamma\text{ap}\rangle \tag{12.12}$$

with
$$\mu(\gamma a,\vec{r})\cdot R_{\gamma a}^{1\alpha}(\vec{r})=\sum_{g \in G}\langle\text{st.}\gamma\text{ap}|g\vec{r}\rangle\langle g\vec{r}|1\alpha\text{ap}\rangle = \frac{\text{ordG}}{\text{dima}}\cdot\sum_{p}\langle\text{st.}\gamma\text{ap}|\vec{r}\rangle\langle\vec{r}|1\alpha\text{ap}\rangle \tag{12.13}$$

This simplifies many relations containing spherical harmonics, for instance the determination of the expansion coefficients (3.22). Inserting (12.12) into (3.22) yields, because of $R_{\gamma a}^{1\alpha}(\vec{S}_i)=R_{\gamma a}^{1\alpha}(\vec{S}_k)$:

$$c(S\beta a,1\alpha) = \sum_{\gamma=1}^{\text{dima}} R_{\gamma a}^{1\alpha}(\vec{S}_k)\cdot\bar{c}(S\beta a,\text{st.}\gamma) \tag{12.14}$$

with an arbitrary $\vec{S}_k \in S$ and a limited set of standard coefficients:

$$\bar{c}(S\beta a, st.\gamma) = \sum_i (S\beta ap|\vec{s}_i)(\vec{s}_i|st.\gamma ap) \qquad (12.15)$$

This reduces the infinite number of coefficients $c(S\beta a, l\alpha)$ belonging to special sets S to a limited number of coefficients defined by (12.15). The scalar functions $R^{l\alpha}_{\gamma a}(\vec{r})$ have to be determined by (12.13). But this has to be done only once for each symmetry group and not for all the sets S in each molecular framework.

In the determination of the SALC coefficients, an arbitrary, unitary transformation (3.23) has been left in obeyance. This problem is posed anew for each polyhedral structure and each equivalent set of vectors, triangles, or tetrahedra. The choice of a standard basis regulates the multiplicity problem in a uniform way for all structures and all equivalent sets within a given symmetry. All SALC coefficients - and therefore all TSALC and QSALC coefficients, too - can be defined via the standard functions.

The proceeding is as follows: Inserting the edge vectors of an equivalent set S into the standard function yields in dima-n(S,a) cases $(\vec{s}_i|S\beta ap)=0$ and in n(S,a) cases $(\vec{s}_i|S\alpha ap)\neq 0$. The latter are necessary and sufficient to determine the SALC coefficients and to fix especially the index α:

$$(\vec{s}_r|S\alpha ap) = (\vec{s}_r|st.\alpha ap)\sqrt{ordG/Z(S)\mu(\alpha a, \vec{s}_r)} \qquad (12.16)$$

Because of the transformation properties of the standard functions, and because of (12.3 to 6), the relations defining the SALC coefficients in section 3 are complied with. By the choice (12.16), the multiplicity indices of all SALC, TSALC, and QSALC coefficients of all molecules sharing one symmetry group are fixed.

By (12.16), the expansion coefficients (12.15) are simplified. Because of (12.3), we get:

$$\bar{c}(S\beta a, st.\gamma) = \delta(\beta,\gamma)\sqrt{\mu(\beta a, \vec{s}_k)Z(S)/ordG} \qquad (12.17)$$

Also the expansion coefficients (3.22) are simplified. Inserting (12.17) into (12.14) yields:

$$c(S\beta a, l\alpha) = R^{l\alpha}_{\beta a}(\vec{s}_k)\sqrt{\mu(\beta a, \vec{s}_k)Z(S)/ordG} \qquad (12.18)$$

In analogy to the parity of the spherical harmonics, $(-\vec{r}|jm)$ $=(-1)^j(\vec{r}|jm)$, we finally define a parity $\{\alpha a\}=\pm 1$ of the standard functions by:

$$(-\vec{r}|st.\alpha ap) = \{\alpha a\}(\vec{r}|st.\alpha ap) \qquad (12.19)$$

From both relations follows:

$$R^{l\alpha}_{\beta a}(-\vec{r}) = (-1)^l\{\beta a\}\cdot R^{l\alpha}_{\beta a}(\vec{r}) \qquad (12.20)$$

13. Transformation properties and structure of the multi-centre
    integrals

13.1. General considerations

The theorems (4.4), (8.1), and (11.2/3) demonstrate that the physical
informations of the multi-centre integrals are concentrated in the invariants BRM, TRM, and QRM. The theorem are proofs of the existence
of these invariants without regard to special atomic orbitals. The theorems (5.5), (8.14), and (11.13) then show, how these informations enter
the reduced (and thereby all ordinary) matrix elements of the molecular
orbitals and geminals, i.e. the global molecular invariants. This relationship is conditioned exclusively by symmetry and geometry.

The functional type of the atomic orbitals (GTO or STO for instance)
affects only the values and functional form of the BRM, TRM, and QRM.
By inversion, as for instance given in (4.10) and (8.7), these can be
determined in principle from given integral formulae. Since this for-
mal inversion requires all multi-centre integrals, it makes sense only if the integral formulae allow to eliminate the quantum number of
the individual atomic orbitals, i.e. the components of the representations. This especially applies to the magnetic quantum numbers.

The general structure of the integrals necessary for this purpose
results from their transformation properties in space. This structure
is not related to a special molecular symmetry. It requires the integral formulae to be tensorial equations with respect to the angular
momentum algebra [15]. Consequently, we have to start with spherical,
atomic orbitals

$$\langle \vec{r} | nlm \rangle = R_{nl}(r) \langle \vec{r} | lm \rangle, \tag{13.1}$$

where the spherical harmonics are defined by (3.18). And further we
have to express the multi-centre integrals by rotational invariants,
3jm symbols of the rotation group O(3), and spherical harmonics of the
atomic distances. Since the rotation group is a supergroup of all
point groups, all the integral formulae can be systematically adapted
to the special molecular symmetry. Thus without further considerations
the polycentric, reduced matrix elements prove to be composed of the
following constituents only: 1) the rotational invariants, 2) the isoscalar factors resulting from the group chain $O(3) \supset G$, 3) the expansion coefficients (3.22) or (12.15) resulting from the spherical harmonics of the atomic distances, and 4) the nj symbols of the concerned
groups O(3) and G. Using atomic spin orbitals we have to refer to
SU(2) and G' instead. The rotational invariants again are the only
carriers of the physical information. They are the only factors de-
ponding on the special, radial functions $R_{nl}(r)$. The other constituents
enter the mediating geometrical factors.

We now derive the universal integral theorems for the atomic orbitals

$$\langle \vec{r} | A n_a l_a m_a \rangle = R_{n_a l_a}(|\vec{r}-\vec{A}|) \cdot \langle \vec{r}-\vec{A} | l_a m_a \rangle \qquad (13.2)$$

defined with respect to the angular momentum basis. These theorems result from the transformation properties.

13.2. Two-centre integrals

The theorem concerning the two-centre integrals states that these integrals over arbitrary, spherical orbitals can be factorized as follows:

$$\langle A n_a l_a m_a | T_M^L | B n_b l_b m_b \rangle \qquad (13.3)$$

$$= \sum_{jJ} \sqrt{4\pi(2J+1)} \cdot \langle n_a l_a \| AB^j, T^L \| n_b l_b \rangle^J \cdot \binom{j \ L \ J}{m \ M \ M}^{+} \binom{1 \ J \ 1}{m_a M' \ m_b}^{+} \langle \vec{AB} | \text{sol } jm \rangle$$

This theorem is the generalization of the Wigner-Eckart theorem. The graphical arrangement of the symbols in the generalized, reduced matrix element shall indicate the coupling of the angular momenta. AB^j stands for the translation operator. The introduction of the solid harmonics

$$\langle \vec{r} | \text{sol } jm \rangle = r^j \langle \vec{r} | jm \rangle \qquad (13.4)$$

ensures the regular behavior for $\vec{A}=\vec{B}$ [44]. If $\vec{A}=\vec{B}$, all terms vanish except for the ordinary, reduced matrix element of the WET:

$$\langle n_a l_a \| 0^0, T^L \| n_b l_b \rangle^L = \langle n_a l_a \| T^L \| n_b l_b \rangle \qquad (13.5)$$

The angular momenta, of course, can be coupled in an order different from that in (13.3). The invariants of the other couplings are related to those of eq.(13.3) by 6j symbols. But only the symmetric arrangement of (13.3) yields the simple conjugation relation:

$$\langle n_b l_b \| BA^j, T^L \| n_a l_a \rangle^J = \langle n_a l_a \| AB^j, T^L \| n_b l_b \rangle^J \cdot (-1)^{j+J+l_a+l_b} \qquad (13.6)$$

In order to prove (13.3), we form the expression

$$I_m^j(\vec{AB}) = \sum \binom{j \ L \ J}{m \ M \ M}^{+} \binom{1 \ J \ 1}{m_a M' \ m_b}^{+} \langle A n_a l_a m_a | T_M^L | B n_b l_b m_b \rangle.$$

It is invariant to translations and therefore a function of the distance vector AB only. We further show the transformation property with respect to rotations:

$$I_m^j(g^{-1}\vec{AB}) = \sum_m D_{mm}^j(g) I_m^j(\vec{AB})$$

Since the spherical harmonics are complete, it follows:

$$I_m^j(\vec{AB}) \sim \langle \vec{AB} | jm \rangle$$

The invariant factor in this proportion, depending on $|\vec{AB}|$ only, is the reduced matrix element. For reasons mentioned above, it is convenient to splitt off the factor $|\vec{AB}|^j$. Converting the result by the orthogonality relations of the 3jm symbols finally yields (13.3).

Examples of (13.3) concerning the identity, the kinetic energy, and
the momentum operator (i.e. L=0 and 1) can be found in [45, 46], and
further integrals of this type with L=0 in [47, 48], cf.also (A2.13).

13.3. Three-centre nuclear attraction integrals

The description of the three-centre integrals is complicated by the
following dilemma: The theorem can be formulated in two different ver-
sions, the more effective of which being at the same time the more dif-
ficult. Because of the translation invariance, only the internal coordinates of the concerned triangle can appear; but there remain many
possibilities of choosing distances and internal angles. Furthermore
different choices may be appropriate to different radial functions,
GTOs or STOs for instance.

Since symmetry considerations suggest an equivalent treatment of
both orbitals, the foolowing reference vectors appear suitable to the
integral $\langle An_a l_a m_a | r_C^{-1} | Bn_b l_b m_b \rangle$:

a) $\vec{AC}$ and $\vec{BC}$

b) $\vec{AB}$ and $\vec{PC}$ with the weighted mean $\vec{P} = (\sigma_A^2 \vec{A} + \sigma_B^2 \vec{B})/(\sigma_A^2 + \sigma_B^2)$

The weight factors σ_X^2 (in the case of GTOs being related to the orbital exponents) may depend on the set X, but not on the individual atomic orbital. Otherwise complications impairing the symmetry considerations arise in section 15. In the case of individual orbital exponents
the vectors AC and BC are preferable.

Since now solid harmonics depending on several distances occur, it
is convenient to introduce the following combinations of solid harmonics:

$$\langle \vec{r}_1, \vec{r}_2 | \text{sol}(j_1, j_2) JM \rangle = \sqrt{2J+1} \sum_M \binom{J \quad j_1 \quad j_2}{M \quad m_1 \quad m_2} \langle \vec{r}_1 | \text{sol } j_1 m_1 \rangle \langle \vec{r}_2 | \text{sol } j_2 m_2 \rangle \quad (13.7)$$

At first, we formulate a weak theorem: Every three-centre integral
over arbitrary, spherical orbitals has the following structure:

$$\langle An_a l_a m_a | r_C^{-1} | Bn_b l_b m_b \rangle$$
$$= 4\pi \sum_{LJj} \sqrt{2L+1} \langle An_a l_a \| AB^J PC^j \| Bn_b l_b \rangle^L \binom{1 \quad 1 \quad L}{m_a \quad m_b \quad M} \langle \vec{AB}, \vec{PC} | \text{sol}(Jj)LM \rangle \quad (13.8)$$

The theorem is weak, because nothing is said about the range of the
summations - except for the triangular conditions for the angular momenta - and nothing about the dependence of the rotational invariants.
In general these may depend on all the scalar variables $|\vec{AB}|$, $|\vec{PC}|$,
and $\vec{AB} \cdot \vec{PC}$. The theorem, of course, applies to any other scalar functions of r_C.

The formulation using the other reference vectors is given by:

$$\langle An_a l_a m_a | r_C^{-1} | Bn_b l_b m_b \rangle$$
$$= 4\pi \sum_{LlI'} \sqrt{2L+1} \langle An_a l_a \| AC^l BC^{I'} \| Bn_b l_b \rangle^L \binom{1 \quad 1 \quad L}{m_a \quad m_b \quad M} \langle \vec{AC}, \vec{BC} | \text{sol}(ll')LM \rangle , \quad (13.9)$$

where the rotational invariants now may be functions of $|\overrightarrow{AC}|$, $|\overrightarrow{BC}|$, and $\overrightarrow{AC} \cdot \overrightarrow{BC}$. The theorems (13.8 and 9) can be converted into one another, since the invariants are interrelated by

$$\langle An_a l_a \| AC^l BC^{l'} \| Bn_b l_b \rangle^L = (2L+1)^{-1} \sum_{Jjnn'} [n l n' l' L \| -\varphi_{AB} \| 0J0jL] \cdot \sigma_A^{2n+1} \sigma_B^{2n'+1'}$$
$$\cdot \theta_{AB}^{-j} \cdot \xi_{AB}^{-J} \cdot AB^{2n} PC^{2n'} \langle An_a l_a \| AB^J PC^j \| Bn_b l_b \rangle^L \qquad (13.10)$$

with $n' = (J+j-l-1)/2 -n$ and conversely by

$$\langle An_a l_a \| AB^J PC^j \| Bn_b l_b \rangle^L = (2L+1)^{-1} \sum_{llNn} [N J n j L \| \varphi_{AB} \| 0l0lL] \cdot \sigma_A^{-1} \cdot \sigma_B^{-1'}$$
$$\cdot \xi_{AB}^{2N+J} \cdot \theta_{AB}^{2n+j} \cdot AC^{2N} BC^{2n} \cdot \langle An_a l_a \| AC^l BC^{l'} \| Bn_b l_b \rangle^L \qquad (13.11)$$

with $n = (1+l'-J-j)/2 -N$. In these equations we have used the abbreviations $\theta_{AB} = \sqrt{\sigma_A^2 + \sigma_B^2}$, $\xi_{AB} = \sigma_A \sigma_B / \theta_{AB}$, and $\varphi_{AB} = \arctan(\sigma_A/\sigma_B)$. The coefficients $[n_3 l_3 n_4 l_4 L \| \varphi \| n_1 l_1 n_2 l_2 L]$ are, except for a different normalization, equal to the Moshinsky-Smirnov coefficients of the Talmi transformation. We refer to [45], eq.(3.4) and the comprehensive references therein.

In order to prove (13.8), we form again:

$$I_M^L(\overrightarrow{AB}, \overrightarrow{PC}) = \sum_{m_a m_b M} \begin{pmatrix} l & l^+ & L^+ \\ m_a & m_b & M \end{pmatrix} \langle An_a l_a m_a | r_C^{-1} | Bn_b l_b m_b \rangle$$

Because of the translational invariance, this is a function of the distances $\overrightarrow{AB}$ and $\overrightarrow{PC}$ only. The rotation property is given by:

$$I_M^L(g^{-1}\overrightarrow{AB}, g^{-1}\overrightarrow{PC}) = \sum_M D_{MM}^L(g) \cdot I_M^L(\overrightarrow{AB}, \overrightarrow{PC})$$

We therefore can express I_M^L by the spherical harmonics of AB and PC or equivalently by the functions defined in (13.7). The expansion coefficients must be scalar, i.e. functions of $|\overrightarrow{AB}|$, $|\overrightarrow{PC}|$, and $\overrightarrow{AB} \cdot \overrightarrow{PC}$ only:

$$I_M^L(\overrightarrow{AB}, \overrightarrow{PC}) = (4\pi\sqrt{2L+1})^{-1} \sum_{Jj} \langle An_a l_a \| AB^J PC^j \| Bn_b l_b \rangle^L \langle \overrightarrow{AB}, \overrightarrow{PC} | \mathrm{sol}(Jj)LM \rangle$$

Converting this equation then yields (13.8). The proof of (13.9) is analogous. The interrelations (13.10/11) result from the following theorem of the solid harmonics defined in (13.7):

$$\langle \vec{r}_1, \vec{r}_2 | \mathrm{sol}(l_1 l_2)LM \rangle = \sum_{n_3 l_3 n_4 l_4} [n_3 l_3 n_4 l_4 L \| \varphi \| 0l_1 0l_2 L] (2L+1)^{-1}$$
$$\cdot r_3^{2n_3} \cdot r_4^{2n_4} \cdot \langle \vec{r}_3, \vec{r}_4 | \mathrm{sol}(l_3 l_4)LM \rangle$$

with $\vec{r}_3 = \vec{r}_1 \cos\varphi - \vec{r}_2 \sin\varphi$ and $\vec{r}_4 = \vec{r}_1 \sin\varphi + \vec{r}_2 \cos\varphi$. This relation is a special case of the more general theorem (3.3) in [45]. If we put $\vec{r}_1 = \xi_{AB}\overrightarrow{AB}$, $\vec{r}_2 = \theta_{AB}\overrightarrow{PC}$, $\varphi = -\varphi_{AB}$, it follows $\vec{r}_3 = \sigma_A\overrightarrow{AC}$, $\vec{r}_4 = \sigma_B\overrightarrow{BC}$ and finally:

$$\langle \overrightarrow{AB}, \overrightarrow{PC} | \mathrm{sol}(l_1 l_2)LM \rangle = \sum_{n_3 l_3 n_4 l_4} [n_3 l_3 n_4 l_4 L \| -\varphi_{AB} \| 0l_1 0l_2 L] (2L+1)^{-1}$$
$$\cdot \sigma_A^{2n_3+l_3} \cdot \sigma_B^{2n_4+l_4} \cdot \xi_{AB}^{-l_1} \cdot \theta_{AB}^{-l_2} \cdot AC^{2n_3} \cdot BC^{2n_4} \cdot \langle \overrightarrow{AC}, \overrightarrow{BC} | \mathrm{sol}(l_3 l_4)LM \rangle$$

Inserting this into (13.8) yields (13.9 and 10). The inverse relation results, if we put $\vec{r}_1 = \sigma_A\overrightarrow{AC}$, $\vec{r}_2 = \sigma_B\overrightarrow{BC}$, and $\varphi = +\varphi_{AB}$.

The weak theorem now can be strengthened in two conflicting aspects by restricting either the dependence of the invariants or the range of the angular momenta J and j (l and l' respectively). The strong theorems are:

A) Every three-centre integral of arbitrary, spherical orbitals has the structure (13.8/9), where the rotational invariants depend on AB and PC (AC and BC) only. This in general causes unlimited sums for J and j (l and l').

B) Every three-centre integral of arbitrary, spherical orbitals has the structure (13.8/9) with the limitation $J+j = l_a+l_b$ ($l+l' = l_a+l_b$). This in general makes the rotational invariants depending on the internal angle, i.e. $\overrightarrow{AB}\cdot\overrightarrow{PC}$ ($\overrightarrow{AC}\cdot\overrightarrow{BC}$).

The structure stated in version B) is preserved by the transformations (13.10/11) in contrast to that of version A).

The integrals of Gauss-type or related orbitals [45-47] occupy a special position. With respect to the reference scheme AB-PC the inequality $J+j \leqslant 2n_a+l_a+2n_b+l_b$ holds for all intergrals. Thus the integrals with $n_a=n_b=0$ are covered by both A) and B).

From the tensor algebraic point of view, the theorem B) has the definite advantage of representing a finite number of integrals by a finite number of invariants, too. Generally in case A), the number is infinite. The limited summation in the case of GTOs is of no much profit, since the group theoretically irrelevant radial quantum numbers interfere in the angular momentum algebra.

On the other hand, explicit formulae of type B) are hard to derive (cf. below), whereas those of type A) result quite naturally from orthogonal expansions. For instance the expansion

$$\langle An_al_am_a|r_C^{-1}|Bn_bl_bm_b\rangle$$

$$= \sum_{n_cn'_c}\sum_{l_cm_c}\langle An_al_am_a|Cn_cl_cm_c\rangle\langle Cn_cl_cm_c|r_C^{-1}|Cn'_cl_cm_c\rangle\langle Cn'_cl_cm_c|Bn_bl_bm_b\rangle$$

in combination with (13.3) yields (13.9) with the invariants

$$\langle An_al_a\|AC^lBC^{l'}\|Bn_bl_b\rangle^L = (-1)^{L+l'+l_a+l_b}\cdot\sum_{n_cn'_cl_c}\sum\sqrt{1/(2l_c+1)}\begin{Bmatrix}L&l&l'\\l_c&l_b&l_a\end{Bmatrix}$$

$$\cdot\langle n_al_a\|AC^l\|n_cl_c\rangle^l\langle n_cl_c\|r^{-1}\|n_cl_c\rangle\langle n_cl_c\|CB^{l'}\|n_bl_b\rangle^{l'}$$

A similar proof follows from the addition theorems discussed in [49]:

$$g(|\vec{r}_1+\vec{r}_2|)\langle\vec{r}_1+\vec{r}_2|LM\rangle$$

$$= \sum_{l_1l_2}(-1)^{-l_1+l_2-L}\sqrt{2L+1}(\begin{smallmatrix}l_1&+l_2&+L\\m&m&M\end{smallmatrix})g_{l_1l_2}^L(\vec{r}_1,\vec{r}_2)\langle\vec{r}_1|l_1m_1\rangle\langle\vec{r}_2|l_2m_2\rangle$$

From this relation follows eq.(5.12) of [49] having the same structure as (13.8/9) except for different reference vectors.

The first proof of B) given in [50] is recursive and thus quite complicated. The results of [46] now allow a simpler proof for a special, complete system of orbitals. Since every orbital can be expanded in such a complete system, the theorem B) is valid for all orbital systems in the Hilbert space. Since the properties of a special orbital system do not matter to the present context, we postpone the proof into the appendix 2.

13.4. The four-centre integrals

There is an even larger variety of representing the four-centre interaction integrals. But the principles are the same as for the three-centre integrals and we can be brief. We confine the representation of the distorted tetrahedra ABCD to the vector triple $\overrightarrow{AC}$, $\overrightarrow{BD}$, and $\overrightarrow{PQ}$ with $\overrightarrow{P} = (\sigma_A^2\overrightarrow{A}+\sigma_C^2\overrightarrow{C})/(\sigma_A^2+\sigma_C^2)$ and $\overrightarrow{Q} = (\sigma_B^2\overrightarrow{B}+\sigma_D^2\overrightarrow{D})/(\sigma_B^2+\sigma_D^2)$. Since now three vectors are involved, we extend (13.7) and define:

$$\langle\vec{r}_1,\vec{r}_2,\vec{r}_3|\text{sol}(j_1j_2)Jj_3LM\rangle$$
$$= \sqrt{2L+1}\cdot\sum\begin{pmatrix}L & J^+ & j_3^+ \\ M & M & m_3\end{pmatrix}\langle\vec{r}_1,\vec{r}_2|\text{sol}(j_1j_2)JM\rangle\langle\vec{r}_3|\text{sol } j_3m_3\rangle \qquad (13.12)$$

Again we first state, in a weak theorem that, the four-centre integrals over arbitrary spherical orbitals pricipally have the following structure:

$$\langle An_al_am_a,Bn_bl_bm_b|r_{12}^{-1}|Cn_cl_cm_c,Dn_dl_dm_d\rangle = \sqrt{4\pi}^3\sum_{l'l''LJj_1}\sum\sqrt{(2l+1)(2l+1)(2L+1)}$$

$$\cdot\left[(An_al_a^+,Cn_cl_c)l\|(AC^{j_4}BD^{j_2})^J{}_{PQ}{}^{j_3}\|(Bn_bl_b^+,Dn_dl_d)l\right]^L \qquad (13.13)$$

$$\cdot\begin{pmatrix}1^+l_ac^1 \\ m_am_cm\end{pmatrix}\begin{pmatrix}1^+l_bd_l' \\ m_bm_dm'\end{pmatrix}\begin{pmatrix}1^+1'^+L^+ \\ m & m' & M\end{pmatrix}\langle\overrightarrow{AC},\overrightarrow{BD},\overrightarrow{PQ}|\text{sol}(j_1 j_2)Jj_3LM\rangle$$

The rotational invariants $[...]^L$ are designed in analogy to the QRMs in (11.2/3) and the reduced matrix elements in (11.9), cf. also [17] eq.(7). In general they are functions of the lengths and the scalar products of the vectors $\overrightarrow{AC}$, $\overrightarrow{BD}$, and $\overrightarrow{PQ}$. Of course, the angular momenta involved can be coupled in a different order. The proof is analogous to that of (13.8/9).

The strong theorems again result from the following restrictions:

A) Every four-centre integral over arbitrary, spherical orbitals has the structure (13.13), where the rotational invariants depend on AC, BD, and PQ only. This in general causes unlimited sums for j_1, j_2, and j_3.

B) Every four-centre integral over arbitrary, spherical orbitals has the structure (13.13) with the limitation $j_1+j_2+j_3 \leq l_a+l_b+l_c+l_d$. This in general makes the rotational invariants depending on the scalar products $\overrightarrow{AC}\cdot\overrightarrow{BD}$, $\overrightarrow{AC}\cdot\overrightarrow{PQ}$, and $\overrightarrow{BD}\cdot\overrightarrow{PQ}$.

Again the GTOs occupy a special position. In the case of this system, the formulae of type A) are restricted by $j_1+j_2+j_3 \leqslant 2n_a+l_a+2n_b+l_b+2n_c$ $+l_c+2n_d+l_d$, cf. [45-47]. Thus the integrals with $n_a=n_b=n_c=n_d=0$ are covered by both theorems A) and B).

The proof of A) can be achieved by orthogonal expansions with respect to the centres P and Q or by the addition theorems already mentioned on page 55. The eq.(5.18) of [49] corresponds to (13.13) with other other reference vectors and another coupling of the angular momenta. The proof of theorem B) is sketched in appendix 2.

In the following, we shall regard the rotational invariants as known. Examples are given in the appendix 2 and in section 23.3/4. But it must be said that the theory of the invariants according to the theorem B) is still unsatisfactory and requires further investigation.

14. Multi-centre integrals and molecular symmetry
14.1. Preliminary considerations

If the symmetry is reduced from the rotation group O(3) to a molecular symmetry group G, the WET is valid with respect to both groups
and the both reduced matrix elements are interrelated by (2.82). We
repeat this relation for O(3):

$$\langle n l m_1 | T_M^L | n' j m_j \rangle = \langle n l \| T^L \| n' j \rangle \begin{pmatrix} 1 & {}^{+}L & j \\ m_1 & M & m_j \end{pmatrix} \qquad (14.1)$$

With the definition

$$| \varphi d p_d \rangle = | n l \delta d p_d \rangle = \sum \langle 1 m_1 | 1 \delta d p_d \rangle \cdot | n l m_1 \rangle \qquad (14.2)$$

the WET reads in G:

$$\langle \varphi d p_d | T_{p_c}^{L \gamma c} | \varphi e p_e \rangle = \sum_\sigma \langle \varphi d \| T^{L \gamma c} \| \varphi e \rangle_\sigma \cdot \begin{pmatrix} d & {}^{+}c & e \\ p_d & p_c & p_e \end{pmatrix}^\sigma \qquad (14.3)$$

where φ has the meaning $\varphi = (n l \delta)$. (2.82) then reads in this case:

$$\langle \varphi d \| T^{L \gamma c} \| \varphi e \rangle_\sigma = \langle n l \| T^L \| n' j \rangle \cdot Is \begin{pmatrix} 1 & {}^{+}L & j \\ \delta & \gamma & \epsilon \\ d & {}^{+}c & e \end{pmatrix}_\sigma \qquad (14.4)$$

Exactly corresponding relations must exist for the structural formulae of the multi-centre integrals discussed in the preceeding section
as generalizations of the WET. These relations are needed, if the universal integral formulae are introduced into the calculations of symmetric molecules.

As a preliminary, we prepare the point group analogues of the spherical harmonics and their combinations (13.7 and 12). As a consequence
of section 12, the standard functions $\langle \vec{r} | st.\alpha a p \rangle$ offer themselves. In
correspondence to (13.7/12), we define the standard functions of seve-
ral variables:

$$\langle \vec{r}_1, \vec{r}_2 | st.(\alpha_1 a_1 \alpha_2 a_2) \beta b p_b \rangle$$
$$= \sqrt{\dim b} \cdot \sum \begin{pmatrix} b & a_1^+ a_2^+ \\ p_b p_1 p_2 \end{pmatrix}^\beta \langle \vec{r}_1 | st.\alpha_1 a_1 p_1 \rangle \langle \vec{r}_2 | st.\alpha_2 a_2 p_2 \rangle \qquad (14.5)$$

$$\langle \vec{r}_1, \vec{r}_2, \vec{r}_3 | st.(\alpha_1 a_1 \alpha_2 a_2) \beta b \alpha_3 a_3 \gamma c p_c \rangle \qquad (14.6)$$
$$= \sqrt{\dim c} \cdot \sum \begin{pmatrix} c & b^+ a_3^+ \\ p_c p_b p_3 \end{pmatrix}^\gamma \langle \vec{r}_1, \vec{r}_2 | st.(\alpha_1 a_1 \alpha_2 a_2) \beta b p_b \rangle \langle \vec{r}_3 | st.\alpha_3 a_3 p_3 \rangle$$

According to (12.12), the s.-a. solid harmonics can be expanded in
standard functions:

$$\langle \vec{r} | sol \ j \sigma a p \rangle = \sum_\alpha S_{\alpha a}^{j \sigma}(\vec{r}) \langle \vec{r} | st.\alpha a p \rangle \quad \text{with } S_{\alpha a}^{j \sigma}(\vec{r}) = r^j \cdot R_{\alpha a}^{j \sigma}(\vec{r}) \qquad (14.7)$$

The corresponding expansions of the functions (13.7) is:

$$\langle \vec{r}_1, \vec{r}_2 | sol(j_1 j_2) J \tau b p_b \rangle = \sqrt{(2J+1)/\dim b} \cdot \sum_{\alpha_1 \sigma_1 a_1 \beta} \sum Is \begin{pmatrix} J & j_1^+ j_2^+ \\ \tau & \sigma_1 \sigma_2 \\ b & a_1^+ a_2^+ \end{pmatrix}_\beta$$
$$\cdot S_{\alpha_1 a_1}^{j_1 \sigma_1}(\vec{r}_1) S_{\alpha_2 a_2}^{j_2 \sigma_2}(\vec{r}_2) \langle \vec{r}_1, \vec{r}_2 | st.(\alpha_1 a_1 \alpha_2 a_2) \beta b p_b \rangle \qquad (14.8)$$

and that of the functions (13.12):

$$\langle \vec{r}_1, \vec{r}_2, \vec{r}_3 | sol(j_1 j_2) J j_3 L \pi c p_c \rangle$$

$$= \sum_{\alpha_1 \sigma_1 a_1 \tau \beta \gamma b} \sum \sqrt{(2L+1)(2J+1)/\text{dimcdimb}} \cdot Is \begin{pmatrix} J & j^+ j_2^+ \\ \tau & \sigma_1 \sigma_2 \\ b & a_1 a_2 \end{pmatrix}_\beta \cdot Is \begin{pmatrix} L & J^+ j_3^+ \\ \pi & \tau \sigma_3 \\ c & b^+ a_3 \end{pmatrix}_\gamma \quad (14.9)$$

$$\cdot S_{\alpha_1 a_1}^{j_1 \sigma_1}(\vec{r}_1) S_{\alpha_2 a_2}^{j_2 \sigma_2}(\vec{r}_2) S_{\alpha_3 a_3}^{j_3 \sigma_3}(\vec{r}_3) \langle \vec{r}_1, \vec{r}_2, \vec{r}_3 | st.(\alpha_1 a_1 \alpha_2 a_2) \beta b \alpha_3 a_3 \gamma c p_c \rangle$$

14.2. The particular integral formulae

Now we are prepared to discuss the multi-centre integrals with respect to molecular symmetry. In each case there are three steps: The formulae of section 13 correspond to (14.1) and it is our task to write down the multi-centre analogues of (14.3) and of the interrelation (14.4).

The point group analogue of the two-centre integral (13.3) is:

$$\langle A \varphi_a a p_a | T_{p_c}^{L \gamma c} | B \varphi_b b p_b \rangle \quad (14.10)$$

$$= \sum_{\delta d \eta \epsilon e} \sum \sqrt{\text{dimd}} \langle \varphi_a a \| AB^{\epsilon e}, T^{L \gamma c} \| \varphi_b b \rangle_{\delta \eta}^d \cdot \binom{e^+ c \ d^+}{p_e p_c p_d}_\eta \binom{a^+ d \ b}{p_a p_d p_b}_\delta \cdot \langle \overline{AB} | st.\epsilon e p_e \rangle$$

and the interrelation of the invariants:

$$\langle \varphi_a a \| AB^{\epsilon e}, T^{L \gamma c} \| \varphi_b b \rangle_{\delta \eta}^d \quad (14.11)$$

$$= \sum_{J J \sigma \tau} \sqrt{4\pi(2J+1)/\text{dimd}} \cdot Is \begin{pmatrix} j^+ L & J^+ \\ \sigma, \gamma & \tau \\ e^+ c & d^+ \end{pmatrix}_\eta Is \begin{pmatrix} l^+ J & l \\ \alpha, \tau & \beta \\ a^+ d & b \end{pmatrix}_\delta S_{\epsilon e}^{J \sigma}(\overline{AB}) \langle n_a l_a \| AB^J, T^L \| n_b l_b \rangle^J$$

Because of the functions $S_{\epsilon e}^{J \sigma}(\overline{AB})$ the point group invariant is no scalar of the rotation group.

The derivation of these equations is as follows. Because of (14.2), the interrelation of the integrals (14.10) and (13.3) is given by:

$$\langle A \varphi_a a p_a | T_{p_c}^{L \gamma c} | B \varphi_b b p_b \rangle$$

$$= \sum \langle 1_a \alpha a p_a | 1_a m_a \rangle \langle LM | L \gamma c p_c \rangle \langle 1_b m_b | 1_b \beta b m_b \rangle \langle A n_a l_a m_a | T_M^L | B n_b l_b m_b \rangle$$

Using now (2.72/73) and (14.7) and comparing the result with (14.10) we get (14.11).

The point group analogue of the three-centre integral (13.8) is:

$$\langle A \varphi_a a p_a | r_C^{-1} | B \varphi_b b p_b \rangle \quad (14.12)$$

$$= \sum_{\epsilon \gamma c \delta d \eta e} \sum \sqrt{\text{dimc}} \langle A \varphi_a a \| AB^{\delta d} PC^{\eta e} \| B \varphi_b b \rangle_{\epsilon \gamma}^c \binom{a^+ b \ c^+}{p_a p_b p_c}^\epsilon \langle \overline{AB}, \overrightarrow{PC} | st.(\delta d \eta e) \gamma c p_c \rangle$$

and again the interrelation of the invariants:

$$\langle A \varphi_a a \| AB^{\delta d} PC^{\eta e} \| B \varphi_b b \rangle_{\epsilon \gamma}^c = 4\pi \sum_{L J J} \sum_{\mu \sigma \tau} \frac{2L+1}{\text{dimc}} \cdot Is \begin{pmatrix} 1^+ 1_b L^+ \\ \alpha_a \beta \mu \\ a^+ b \ c^+ \end{pmatrix}_\epsilon Is \begin{pmatrix} L & J^+ j^+ \\ \mu & \sigma, \tau \\ a & d^+ e^+ \end{pmatrix}_\gamma \quad (14.13)$$

$$\cdot S_{\delta d}^{J \sigma}(\overline{AB}) S_{\eta e}^{j \tau}(\overrightarrow{PC}) \langle A n_a l_a \| AB^J PC^j \| B n_b l_b \rangle^L$$

These equations again follow from

$$\langle A \varphi_a a p_a | r_C^{-1} | B \varphi_b b p_b \rangle = \sum \langle 1_a \alpha a p_a | 1_a m_a \rangle \langle 1_b m_b | 1_b \beta b p_b \rangle \langle A n_a l_a m_a | r_C^{-1} | B n_b l_b m_b \rangle$$

using the expansion (14.8).

With reference to the vectors $\vec{AC}$ and $\vec{BC}$ the corresponding formulae read:

$$\langle A\varphi_a ap_a | r_C^{-1} | B\varphi_b bp_b \rangle$$

$$= \sum_{\varepsilon\gamma c\varphi f} \sum_{\varphi f \varphi' f'} \sqrt{\text{dimc}} \langle A\varphi_a \| AC^{\varphi f} BC^{\varphi' f'} \| B\varphi_b b \rangle^c_{\varepsilon\gamma} \begin{pmatrix} a^+ b & c^+ \\ p_a p_b p_c \end{pmatrix}^\varepsilon_c \langle \vec{AC}, \vec{BC} | st.(\varphi f \varphi' f')\gamma c p_c \rangle \qquad (14.14)$$

with

$$\langle A\varphi_a \| AC^{\varphi f} BC^{\varphi' f'} \| B\varphi_b b \rangle^c_{\varepsilon\gamma} = 4\pi \sum_{LL'L} \sum_{\mu\pi\pi'} \frac{2L+1}{\text{dimc}} \cdot \text{Is} \begin{pmatrix} 1^+_a & 1_b & L^+ \\ \alpha & \beta & \mu \\ a^+b & c^+ \end{pmatrix}_\varepsilon \text{Is} \begin{pmatrix} L & 1^+ 1'^+ \\ \mu & \pi & \pi' \\ c & f^+ f'^+ \end{pmatrix}_\gamma$$

$$\cdot S^{1\pi}_{\varphi f}(\vec{AC}) S^{1'\pi'}_{\varphi' f'}(\vec{BC}) \langle An_a 1_a \| AC^1 BC^{1'} \| Bn_b 1_b \rangle^L \qquad (14.15)$$

In section 13, we have criticized that the three-centre integral formulae of type A) can contain infinitely many rotational invariants. This problem does not occur in the case of molecular symmetry, since the sums in (14.12 and 14) are finite because of the limited number of point group representations. But in truth the problem has only been shifted, since according to (14.13 and 14) the finite number of point group invariants is expressed by an infinite number of rotational invariants. For type B), of course, this problem does not exist.

In conclusion, we carry out the corresponding determination of the two-electron integrals. The parallelism of rotation and point group yields the analogue of (13.13):

$$\langle A\varphi_a ap_a, B\varphi_b bp_b | r_{12}^{-1} | C\varphi_c cp_c, D\varphi_d dp_d \rangle = \sum_{\eta_1 h_1} \sum_{\sigma\tau g\varepsilon\varepsilon\varepsilon\varepsilon\varphi f} \sum \sqrt{\text{dime}\cdot\text{dime}\cdot\text{dimg}}$$

$$\cdot [(A\varphi_a a^+, C\varphi_c c)\varepsilon e \| (AC^{\eta_4 h_4} BD^{\eta_2 h_2})^{\varphi f} PQ^{\eta_3 h_3} \| (B\varphi_b b^+, D\varphi_d d)\acute{\varepsilon}\acute{e}]^g_{\sigma\tau} \qquad (14.16)$$

$$\cdot \begin{pmatrix} a^+ c & e \\ p_a p_c p_e \end{pmatrix}_\varepsilon \begin{pmatrix} b^+ d & \acute{e} \\ p_b p_d p_e \end{pmatrix}_{\acute{\varepsilon}} \begin{pmatrix} e^+ \acute{e}^+ g \\ p_e p_{\acute{e}} p_g \end{pmatrix}^\sigma_o \langle \vec{AC}, \vec{BD}, \vec{PQ} | st.(\eta_1 h_1 \eta_2 h_2)\varphi f \eta_3 h_3 \tau g p_g \rangle$$

The point group invariants of this theorem are related to the rotational invariants by:

$$[(A\varphi_a a^+, C\varphi_c c)\varepsilon e \| (AC^{\eta_4 h_4} BD^{\eta_2 h_2})^{\varphi f} PQ^{\eta_3 h_3} \| (B\varphi_b b^+, D\varphi_d d)\acute{\varepsilon}\acute{e}]^g_{\sigma\tau}$$

$$= \sqrt{4\pi^3} \cdot \sum_{j_4 \pi_4} \sum_{j\mu L\gamma 1l'1l} \sum_{\text{dimg}} \frac{2L+1}{\text{dimg}} \cdot \sqrt{\frac{(2l+1)(2l'+1)(2J+1)}{\text{dime}\cdot\text{dime}\cdot\text{dimf}}} \cdot \text{Is} \begin{pmatrix} 1^+_a & c & 1 \\ \alpha & \gamma & \lambda \\ a^+ c & e \end{pmatrix}_\varepsilon \text{Is} \begin{pmatrix} 1^+_b & d & 1' \\ \beta & \delta & \lambda' \\ b^+ d & \acute{e} \end{pmatrix}_{\acute{\varepsilon}}$$

$$\cdot \text{Is} \begin{pmatrix} J & j^+_1 j^+_2 \\ \mu & \pi_1 \pi_2 \\ f & h_1 h_2 \end{pmatrix}_\varphi \text{Is} \begin{pmatrix} 1^+ 1'^+ L^+ \\ \lambda \lambda' \gamma \\ e^+ \acute{e}^+ g^+ \end{pmatrix}_\sigma \text{Is} \begin{pmatrix} L & J^+ j^+_3 \\ \gamma & \mu & \pi_3 \\ g & f^+ h_3^+ \end{pmatrix}_\tau \cdot S^{j_4 \pi_4}_{\eta_4 h_4}(\vec{AC}) S^{j_2 \pi_2}_{\eta_2 h_2}(\vec{BD}) S^{j_3 \pi_3}_{\eta_3 h_3}(\vec{PQ}) \qquad (14.17)$$

$$\cdot [(An_a 1^+_a, Cn_c 1_c) 1 \| (AC^{j_4} BD^{j_2})^J PQ^{j_3} \| (Bn_b 1^+_b, Dn_d 1_d) 1']^L$$

This results from (13.13) by the adaption to the point group symmetry (14.2). One then uses (2.72/73) and the expansion (14.9).

15. Ab initio determination of the polycentric invariants

Using the integral formulae of section 13 and their adaption to the particular molecular symmetry in section 14, we now can determine the polycentric, reduced matrix elements BRM, TRM, and QRM.

15.1. The proper two-centre integrals

Going to calculate we have to rewrite (14.10) for the centres $\vec{A}_1$ and $\vec{B}_k$. Noting that $\overrightarrow{A_1 B_k} = \vec{B}_k - \vec{A}_1 = -\vec{S}_{ik}$, we get with (12.19):

$$\langle A i \varphi_a a p_a | T^c_{p_c} | B k \varphi_b b p_b \rangle$$
$$= \sum_{\delta d \eta \epsilon \epsilon'} \sum \sqrt{\dim d} \langle \varphi_a a \| -S^{\epsilon'e}, T^c \| \varphi_b b \rangle^d_{\delta \eta} \cdot \begin{pmatrix} e^+c & d \\ p_e p_c p_d \end{pmatrix}^\eta \begin{pmatrix} a^+d & b \\ p_a p_d p_b \end{pmatrix}^\delta \{ \epsilon' e \} \langle \vec{S}_{ik} | st.\epsilon' e p_e \rangle$$

Inserting this into (4.10) yields the BRMs expressed by the point group invariants:

$$(A \varphi_a a \| T^c \| B \varphi_b b)^{d \delta \eta}_{S \epsilon e}$$
$$= \{ c^+ d e \eta \} \{ a^+ b d \delta \} \sqrt{\dim d} \cdot \sum_{\epsilon'} \bar{c} (S \epsilon e, st.\epsilon') \{ \epsilon' e \} \langle \varphi_a a \| -S^{\epsilon' e}, T^c \| \varphi_b b \rangle^d_{\delta \eta} \qquad (15.1)$$

Using now (14.11) yields the final expression by the rotational invariants of the general two-centre integrals:

$$(A \varphi_a a \| T^{L \gamma c} \| B \varphi_b b)^{d \delta \eta}_{S \epsilon e} = (-1)^{l_a + l_b + L} \sum_{J J \sigma \tau \epsilon'} \sum \bar{c} (S \epsilon e, st.\epsilon') \sqrt{4 \pi (2J+1)} \cdot S^{j \sigma}_{\epsilon' e} (\vec{S}_r)$$
$$\cdot Is \begin{pmatrix} j^+ J^+ L \\ \sigma + \tau + \gamma \\ e + d^+ c \end{pmatrix}_\eta Is \begin{pmatrix} 1^+_a 1_b J \\ \alpha \beta \tau \\ a^+ b \ d \end{pmatrix}_\delta \langle n_a l_a \| -S^J, T^L \| n_b l_b \rangle^J \qquad (15.2)$$

This completes the ab inition calculation of the two-centre matrix elements. In the next step, the matrix elements of the s.-a. MOs (5.1) or (11.6) can be expressed directly by the rotational invariants. The abbreviation λ now includes the main and the angular momentum quantum numbers, i.e. $\lambda = (A \dot{\alpha} \dot{a}', \varphi_a a) \dot{\alpha} = (A \dot{\alpha} \dot{a}', n_a l_a \alpha a) \dot{\alpha}$. The s.-a. MOs are:

$$| \lambda \dot{a} p_a \rangle = \sum_{I q_a} K(\dot{\alpha} \dot{a} p_a, A i \alpha a, a q_a) \cdot | A i \varphi_a a q_a \rangle$$
$$= \sum_{I q_a m_a} \sum K(\dot{\alpha} \dot{a} p_a, A i \dot{a}', a q_a) \langle 1_a m_a | 1_a \alpha a q_a \rangle \cdot | A i n_a l_a m_a \rangle \qquad (15.3)$$
$$= \sum_{I m_a} M(\dot{\alpha} \dot{a} p_a, A i \dot{a}', (\alpha a) l_a m_a) \cdot | A i n_a l_a m_a \rangle$$

The coefficients M(...) relating the s.-a. MOs directly to the spherical AOs have been defined in [51], eq.(7). From the orthogonality relations (2.26/27), (2.70/71), and (3.6/7) follow those of the composed coefficients M:

$$\sum_{I m} M(\dot{\alpha} \dot{a} p_a, A i \dot{a} \dot{a}', (\alpha a) l m) \cdot M(\beta \dot{b} p_b, A i \beta \dot{b}', (\beta b) l m) \qquad (15.4)$$
$$= \delta(\dot{\alpha}, \dot{\beta}) \delta(\dot{a}, \dot{b}) \delta(p_a, p_b) \delta(\dot{a}', \dot{\beta}') \delta(a', b') \delta(\alpha, \beta) \delta(a, b)$$
$$\sum_{\dot{a}' \dot{a} \alpha a \dot{\alpha} p_a} \sum M(\dot{\alpha} \dot{a} p_a, A i \dot{a} \dot{a}', (\alpha a) l m) \cdot M(\dot{\alpha} \dot{a} p_a, A j \dot{a} \dot{a}', (\alpha a) l n) = \delta(i, j) \delta(m, n) \qquad (15.5)$$

The matrix elements of the MOs (15.3) thus are given by the WET

$$\langle \mathring{A}\dot{a}p_a | T_{P_c}^{L\gamma c} | \mathring{B}\dot{b}p_b \rangle = \sum_\mu \langle \mathring{A}\dot{a} || T^{L\gamma c} || \mathring{B}\dot{b} \rangle_\mu \cdot \binom{\mathring{a}^+c\ \mathring{b}}{p_a p_c p_b}\mu \qquad (15.6)$$

with the reduced matrix elements

$$\langle \mathring{A}\dot{a} || T^{L\gamma c} || \mathring{B}\dot{b} \rangle_\mu = \sum_{JjS} \langle n_a 1_a || -S^J, T^L || n_b 1_b \rangle^J \cdot GEO_4(\mathring{A}\dot{a}, \mathring{B}\dot{b}, L\gamma c\mu, JjS) \qquad (15.7)$$

and the geometrical factor:

$$GEO_4(\mathring{A}\dot{a}, \mathring{B}\dot{b}, L\gamma c\mu, JjS) = (-1)^{1_a+1_b L} \cdot \sqrt{4\pi(2J+1)} \cdot \sum_{\delta\eta\epsilon\epsilon} \sum_{ste'} \bar{c}(See, st.\dot{\epsilon}')$$

$$\cdot S_{\dot{\epsilon}e}^{j\sigma}(\vec{S}_r) \cdot Is \begin{pmatrix} j^+ J^+ L \\ \sigma, \tau, \gamma \\ e^+ d^+ c \end{pmatrix}_\eta Is \begin{pmatrix} 1_a^+ J \\ \alpha_a^+ \beta_b^+ \tau \\ a^+ b\ d \end{pmatrix}_\delta GEO_1(AaBbc, \dot{a}'\dot{a}\dot{a}\beta'\dot{b}\dot{b}\mu, \delta\eta See) \qquad (15.8)$$

The paper [51] dealt with the overlap matrix, i.e. the matrix elements of the identity operator, a special case of (15.6-8).

15.2. The nuclear attraction integrals

For the determination of the TRMs, we have to repeat (14.12 or 14) with respect to the centres $\vec{A}_i$, $\vec{B}_k$, and $\vec{C}_l$. This yields an invariant $\langle A\varphi_a a || A_i B_k^{\delta d} P_{1k} C_1^{\eta e} || B\varphi_b b \rangle_{\epsilon\gamma}^c$ with $\vec{P}_{1k} = (\sigma_A^2 \vec{A}_i + \sigma_B^2 \vec{B}_k)/(\sigma_A^2 + \sigma_B^2)$. Because of the invariance the vectors $\overrightarrow{A_i B_k}$ and $\overrightarrow{P_{1k} C_1}$ can be replaced by those of an arbitrary, equivalent triangle $\overrightarrow{A_x B_y}$ and $\overrightarrow{P_{xy} C_z}$. In order to eliminate the misleading indices ikl (or xyz) we replace them by $\overrightarrow{AB}_\Delta$ and $\overrightarrow{PC}_\Delta$, where the index Δ indicates the set of the triangles under consideration. Inserting now the specified relation (14.12) into (8.7) then yields:

$$(A\varphi_a a || Cr^{-1} || B\varphi_b b)_{\Delta\gamma c}^\epsilon = \sqrt{dimc} \sum_{\gamma\delta d\eta e} \sum \langle A\varphi_a a || AB_\Delta^{\delta d} PC_\Delta^{\eta e} || B\varphi_b b \rangle_{\epsilon\gamma'}^c \cdot \bar{c}_1^2(\Delta\gamma c, st.\delta d\eta e\gamma') \qquad (15.9)$$

with the expansion coefficients:

$$\bar{c}_1^2(\Delta\gamma c, st.\delta d\eta e\gamma') = \sum_{ijkl}(\Delta\gamma cn | \Delta_j) \cdot \sqrt{Z(\Delta)}\tau\binom{\Delta ACB}{jilk} \langle \overrightarrow{A_i B_k}, \overrightarrow{P_{1k} C_1} | st.(\delta d\eta e)\gamma'cn \rangle \qquad (15.10)$$

These coefficients result from the expansion

$$\langle \overrightarrow{A_i B_k}, \overrightarrow{P_{1k} C_1} | st.(\delta d\eta e)\gamma cn \rangle = \sum_\gamma \bar{c}_1^2(\Delta\gamma c, st.\delta d\eta e\gamma) \sum_{\Delta_j} \sqrt{Z(\Delta)}\tau\binom{\Delta ACB}{jilk}(\Delta_j | \Delta\gamma cn) \qquad (15.11)$$

and are the analogues of (12.15).

On the other hand, the insertion of (14.14) into (8.7) yields:

$$(A\varphi_a a || Cr^{-1} || B\varphi_b b)_{\Delta\gamma c}^\epsilon = \sqrt{dimc} \sum_{\varphi f\varphi' f\gamma'} \sum \langle A\varphi_a a || AC_\Delta^{\varphi f} BC_\Delta^{\varphi' f} || B\varphi_b b \rangle_{\epsilon\gamma'}^c \cdot \bar{c}_2^2(\Delta\gamma c, st.\varphi f\varphi' f\gamma') \qquad (15.12)$$

with a different type of coefficients:

$$\bar{c}_2^2(\Delta\gamma c, st.\varphi f\varphi' f\gamma') = \sum_{ijkl}(\Delta\gamma cn | \Delta_j)\sqrt{Z(\Delta)}\tau\binom{\Delta ACB}{jilk}\langle \overrightarrow{A_i C_1}, \overrightarrow{B_k C_1} | st.(\varphi f\varphi' f)\gamma'cn \rangle \qquad (15.13)$$

The eqs.(15.9 and 12) show that different types of coefficients have to be calculated depending on the reference vectors in the integral formulae. Therefore it is absolutely inexpedient to allow the parameters σ_A and σ_B in the weighted mean $\vec{P}_{1k}$ to depend on the individual orbitals at a certain centre. Problems also arise, if approximative formulae require other reference vectors.

Inserting now (14.13) into (15.9) and (14.15) into (15.12) traces back the TRMs to the rotational invariants. With respect to the reference vectors $\overrightarrow{AC}$ and $\overrightarrow{BC}$, the result is:

$$(A\varphi_a a\|Cr^{-1}\|B\varphi_b b)^\varepsilon_{\Delta\gamma c}$$
$$= \sum_{1\tilde{1}L} GEO_5(Al_a\alpha a, Bl_b\beta b, C\Delta\gamma c\varepsilon, 1\tilde{1}L) \cdot \left\langle An_a l_a\|AC^1_\Delta BC^{1'}_\Delta\| Bn_b l_b\right\rangle^L \tag{15.14}$$

with the geometrical factor:

$$GEO_5(Al_a\alpha a, Bl_b\beta b, C\Delta\gamma c\varepsilon, 1\tilde{1}L) = 4\pi(2L+1)\,\mathrm{dimc}^{-1/2} \sum_{\varphi f\varphi f'\gamma'\mu\pi\pi'} \bar{c}^2_2(\Delta\gamma c, \mathrm{st.}\varphi f\varphi' f'\gamma')$$
$$\cdot Is\begin{pmatrix} 1^+ & 1^+ & L^+ \\ \alpha & \beta & \gamma' \\ a^+ & b & c^+ \end{pmatrix}_\varepsilon Is\begin{pmatrix} L & 1^+ & 1^+ \\ \mu & \pi & \pi' \\ c & f^+ & f'^+ \end{pmatrix}_{\gamma'} S^{1\pi}_{\varphi f}(\overrightarrow{AC}_\Delta) S^{1\pi}_{\varphi' f'}(\overrightarrow{BC}_\Delta) \tag{15.15}$$

Finally the matrix elements of the potential operator (8.10/11) with respect to the s.-a. MOs can be determined:

$$\langle \mathring{A}\mathring{p}_a|V_c|\mathring{B}\mathring{p}_a\rangle = \langle \mathring{A}\mathring{a}\|V_c\|\mathring{B}\mathring{a}\rangle/\sqrt{\mathrm{dim}\mathring{a}} \tag{15.16}$$
$$= \sqrt{1/\mathrm{dim}\mathring{a}} \cdot \sum_{1\tilde{1}L\Delta} GEO_6(\mathring{A}\mathring{a}\mathring{B}\mathring{a}C, 1\tilde{1}L\Delta)\langle An_a l_a\|AC^1_\Delta BC^{1'}_\Delta\|Bn_b l_b\rangle^L$$

with

$$GEO_6(\mathring{A}\mathring{a}\mathring{B}\mathring{a}C, 11L\Delta) = \sum_{\gamma c\varepsilon} GEO_5(Al_a\alpha a, Bl_b\beta b, C\Delta\gamma c\varepsilon, 1\tilde{1}L) \tag{15.17}$$
$$\cdot GEO_2(AaBbC, \mathring{\alpha}\mathring{a}\mathring{\beta}\mathring{b}\mathring{a}\mathring{a}, \Delta\gamma c\varepsilon)$$

15.3. The interaction integrals

In the same way as before, we restate (14.16) for the centres $\overrightarrow{A_i}$, $\overrightarrow{B_j}$, $\overrightarrow{C_k}$, and $\overrightarrow{D_l}$. The resulting integral is equal to (11.1-3) and the isolation of the QRM give the expression in terms of the point group invariants. Again we introduce the vectors $\overrightarrow{AC}_\mathfrak{F}$, $\overrightarrow{BD}_\mathfrak{F}$, and $\overrightarrow{PQ}_\mathfrak{F}$(instead of $\overrightarrow{A_iC_k}$, $\overrightarrow{B_jD_l}$, and $\overrightarrow{P_{ik}Q_{jl}}$) referring to an arbitrary member of the set $\mathfrak{F}$. The QRMs are:

$$\left[(A\varphi_a a^+, C\varphi_c c)\varepsilon e\|r^{-1}_{12}\|(B\varphi_b b^+, D\varphi_d d)e'\acute{e}\right]^\sigma_{\tau g} =$$
$$= \sum_{\eta_1 h_1} \sum_{\tau\varphi f} \sqrt{\mathrm{dimg}} \cdot \bar{c}^3(\mathfrak{F}\tau g, \mathrm{st.}(\eta_1 h_1 \eta_2 h_2)\varphi f\eta_3 h_3 \tau') \tag{15.18}$$
$$\cdot \left[(A\varphi_a a^+, C\varphi_c c)\varepsilon e\|(AC^{\eta_1 h_1}_\mathfrak{F}, BD^{\eta_2 h_2}_\mathfrak{F})\varphi f PQ^{\eta_3 h_3}_\mathfrak{F}\|(B\varphi_b b^+, D\varphi_d d)e'\acute{e}\right]^g_{\sigma\tau}$$

where the expansion coefficients are given by:

$$\bar{c}^3(\mathfrak{F}\tau g, \mathrm{st.}(\eta_1 h_1 \eta_2 h_2)\varphi f\eta_3 h_3 \tau') \tag{15.19}$$
$$= \sum_{rijkl} \sum (\mathfrak{F}\tau g p_g|\tilde{\sum}_r)\sqrt{Z(\mathfrak{F})}\tau\binom{\mathfrak{F}ABCD}{rijkl}\langle \overrightarrow{A_iC_k}, \overrightarrow{B_jD_l}, \overrightarrow{P_{ik}Q_{jl}}|\mathrm{st.}(\eta_1 h_1\eta_2 h_2)\varphi f\eta_3 h_3 \tau' g p_g\rangle$$

These again are obvious generalizations of the coefficients (15.10/13) and (12.15). They likewise depend on the system of reference vectors within the pseudo-tetrahedra.

Inserting now (15.18) into (14.17) traces back the QRMs to the rotational invariants:

$$\left[(A\varphi_a a^+, C\varphi_c c)\varepsilon e \| r_{12}^{-1} \| (B\varphi_b b^+, D\varphi_d d)\acute{e}\acute{e}\right]^{\sigma}_{\mathcal{F}\tau g} \tag{15.20}$$

$$= \sum_{j_i J} \sum_{I\mathcal{I}L} GEO_7(A\varphi_a aC\varphi_c c\varepsilon e, B\varphi_b bD\varphi_d d\acute{e}\acute{e}, \sigma \mathcal{F}\tau g; j_i J(1\mathcal{I}')L)$$

$$\cdot \left[(An_a 1_a^+, Cn_c 1_c)1\|(AC_{\mathcal{F}}^{j_1} BD_{\mathcal{F}}^{j_2})^J PQ_{\mathcal{F}}^{j_3} \|(Bn_b 1_b^+, Dn_d 1_d)1\right]^L$$

with the geometrical factor:

$$GEO_7(...) = (4\pi)^{3/2}((2L+1)/\text{dimg}) \sum_{\eta_i h_i} \sum_{\tau\varphi I} \left[\frac{(2l+1)(2l'+1)(2J+1)}{\text{dime}\cdot\text{dime}'\cdot\text{dimf}}\right]^{1/2} \tag{15.21}$$

$$\cdot \sum_{\pi_i\mu} \sum_{\gamma'\lambda\lambda'} Is\begin{pmatrix}1 & 1 & 1 \\ \alpha^+ & a & c \\ a^+ & \gamma & \lambda \\ c & e\end{pmatrix}_\varepsilon \, Is\begin{pmatrix}1 & 1 & 1' \\ \beta^+ & b & d \\ b^+ & \delta & \lambda' \\ d & e'\end{pmatrix}_{\varepsilon'} \, Is\begin{pmatrix}J & j_1^+ & j_2^+ \\ \mu & \pi_1 & \pi_2 \\ f & h_1 h_1' & h_2\end{pmatrix}_\varphi \, Is\begin{pmatrix}1^+ & 1'^+ & L^+ \\ \lambda & \lambda' & \gamma' \\ e^+ & e'^+ & g^+\end{pmatrix}_\sigma \, Is\begin{pmatrix}L & J^+ & j_3^+ \\ \gamma' & \mu & \pi_3 \\ g & f^+ & h_3\end{pmatrix}_{\tau'}$$

$$\cdot \bar{\sigma}^3(\mathcal{F}\tau g, \text{st.}(\eta_1 h_1 \eta_2 h_2)\varphi f \eta_3 h_3 \tau') \cdot S^{j_1 \pi_1}_{\eta_1 h_1}(\overrightarrow{AC_{\mathcal{F}}}) S^{j_2 \pi_2}_{\eta_2 h_2}(\overrightarrow{BD_{\mathcal{F}}}) S^{j_3 \pi_3}_{\eta_3 h_3}(\overrightarrow{PQ_{\mathcal{F}}})$$

And finally, we determine the two-electron matrix elements of the s.-a. LCAO-MOs. From (11.13) and (15.20) results:

$$\left[(A'\grave{a}^+, \phi\grave{c})\theta t \| r_{12}^{-1} \| (B'b^+, Dd)\acute{\theta}t\right] = \sum_{j_i J} \sum_{I\mathcal{I}L\mathcal{F}} GEO_8(A'\grave{a}\phi\grave{c}\theta, B'b Dd\acute{\theta}t; j_i J(1\mathcal{I}')L\mathcal{F})$$

$$\cdot \left[(An_a 1_a^+, Cn_c 1_c)1\|(AC_{\mathcal{F}}^{j_1} BD_{\mathcal{F}}^{j_2})^J PQ_{\mathcal{F}}^{j_3} \|(Bn_b 1_b^+, Dn_d 1_d)1\right]^L \tag{15.22}$$

with the geometrical factor:

$$GEO_8(A'\grave{a}\phi\grave{c}\theta, B'b Dd\acute{\theta}t; j_i J(1\mathcal{I}')L\mathcal{F}) \tag{15.23}$$

$$= \sum_{\varepsilon e\varepsilon'\acute{e}} \sum_{\sigma\tau g} GEO_3(\varepsilon e^{-j}\acute{d}\sigma, \mathcal{F}\tau g) \cdot GEO_7(A\varphi_a aC\varphi_c c\varepsilon e, B\varphi_b bD\varphi_d d\acute{e}\acute{e}, \sigma\mathcal{F}\tau g; j_i J(1\mathcal{I}')L)$$

16. Wolfsberg-Helmholtz and Mulliken approximations

Having cleared up the interrelation between the molecular invariants defined in the first sections and the ab initio rotational invariants, we now consider approximations like those of Wolfsberg-Helmholtz and Mulliken. These approximations can be discussed from three different, but connected points of view: namely as approximative formulae for distant centres, as a minimization of the number of semi-empirical parameters, and as an interpretational aid to split up the molecular energy into a quasi-classic part and the "remainder" [52].

At first, it is a trivial requirement, that an approximation or parametrization should not depend on the position and orientation of the molecule in space, i.e. that it should be invariant to translations and rotations. This applies in particular to the symmetry operations of the molecule. But the existence of a relevant literature suggests, that this trivial requirement is by no means a matter of fact [53-55].

The required invariance is self-evident, if the approximations are sujected to the principles of form stated in section 13, if especially both sides of the approximative formulae are tensors of the same species. This condition is not met by the usual form of the Wolfsberg-Helmholtz approximation:

$$\langle An_{a}l_{a}m_{a}|r_{C}^{-1}|Bn_{b}l_{b}m_{b}\rangle$$
$$= k\langle An_{a}l_{a}m_{a}|Bn_{b}l_{b}m_{b}\rangle\left[\langle An_{a}l_{a}m_{a}|r_{C}^{-1}|Bn_{a}l_{a}m_{a}\rangle+\langle Bn_{b}l_{b}m_{b}|r_{C}^{-1}|Bn_{b}l_{b}m_{b}\rangle\right]$$

The left side and both summands on the right transform according to three different product representations of the rotation group. This demands a meaningful modification, which for instance can be achieved by taking the mean with respect to the magnetic quantum numbers. There are two possibilities:

$$\langle An_{a}l_{a}m_{a}|r_{C}^{-1}|Bn_{b}l_{b}m_{b}\rangle = \frac{k}{2l_{b}+1}\cdot\sum_{m_{b}'}\langle An_{a}l_{a}m_{a}|Bn_{b}l_{b}m_{b}'\rangle\langle Bn_{b}l_{b}m_{b}'|r_{C}^{-1}|Bn_{b}l_{b}m_{b}\rangle$$
$$(16.1)$$
$$+ \frac{k}{2l_{a}+1}\cdot\sum_{m_{a}'}\langle An_{a}l_{a}m_{a}'|Bn_{b}l_{b}m_{b}\rangle\langle An_{a}l_{a}m_{a}|r_{C}^{-1}|An_{a}l_{a}m_{a}'\rangle$$

or

$$\langle An_{a}l_{a}m_{a}|r_{C}^{-1}|Bn_{b}l_{b}m_{b}\rangle = k\langle An_{a}l_{a}m_{a}|Bn_{b}l_{b}m_{b}\rangle$$
$$(16.2)$$
$$\cdot\left[\frac{1}{2l_{a}+1}\cdot\sum_{m_{a}}\langle An_{a}l_{a}m_{a}|r_{C}^{-1}|An_{a}l_{a}m_{a}\rangle + \frac{1}{2l_{b}+1}\cdot\sum_{m_{b}}\langle Bn_{b}l_{b}m_{b}|r_{C}^{-1}|Bn_{b}l_{b}m_{b}\rangle\right]$$

The first approximation depends on the angles between the vectors $\overrightarrow{AB}$ and $\overrightarrow{AC}$ or $\overrightarrow{BC}$ respectively. The second one is simpler and leads to a smaller number of parameters. Furthermore it fits with the reference vectors AB-PC and we can use the coefficients (15.10). We therefore prefer (16.2) and determine the corresponding approximation of the rotational invariants. From (13.8) follows:

$$\sum_{m_a} \langle An_a l_a m_a | r_C^{-1} | An_a l_a m_a \rangle = \sqrt{2l_a+1} \langle An_a l_a \| 0^\circ AC^\circ \| An_a l_a \rangle^\circ$$

$$= \sqrt{2l_a+1} \cdot \langle n_a l_a \| n_a l_a \rangle$$

with the ordinary, reduced matrix element of (13.5). If we insert this into (16.2) and compare the result with (13.8), we get:

$$\langle An_a l_a \| AB^J PC^J \| Bn_b l_b \rangle^L = \delta(J,0)\delta(J,L)\langle n_a l_a \| AB^L \| n_b l_b \rangle^L (-1)^{L+l_a+l_b}$$

$$\cdot \left[1/\sqrt{2l_a+1} \langle n_a l_a \| n_a l_a \rangle + 1/\sqrt{2l_b+1} \langle n_b l_b \| n_b l_b \rangle \right] \tag{16.3}$$

The approximation with respect to the angular momentum basis is the most efficient and selfconsistent. But a similar approximation is possible on the lower level of the point group invariants (section 14) or even on the level of the TRMs. The latter presents itself, if the parametrization shall not require spherical, atomic orbitals. Approximating (8.1) in the sense of (16.2),

$$\langle Ai\varphi_a am | r_{C_1}^{-1} | Bk\varphi_b bq \rangle = k\langle Ai\varphi_a am | Bk\varphi_b bq \rangle$$

$$\cdot \left[\frac{1}{dima} \sum_m \langle Ai\varphi_a am' | r_{C_1}^{-1} | Ai\varphi_a am \rangle + \frac{1}{dimb} \sum_{q'} \langle Bk\varphi_b bq' | r_{C_1}^{-1} | Bk\varphi_b bq' \rangle \right], \tag{16.4}$$

yields for the TRMs:

$$(A\varphi_a a \| Cr^{-1} \| Bk\varphi_b b)_{\Delta\gamma c}^\varepsilon = k\frac{\sqrt{Z(\Delta)Z(C)}}{dimc} \sum_{S\sigma} (A\varphi_a a \| B\varphi_b b)_{S\sigma c}^\varepsilon \cdot PIs^2 \begin{pmatrix} \Delta & S & C \\ \gamma_+ \varepsilon & 1 \\ c+c & 1 \end{pmatrix} \begin{vmatrix} \\ 1 \\ 1 \end{vmatrix}$$

$$\cdot \sum_{Xx} (1/\sqrt{Z(XCX)dimx}) \cdot (X\varphi_x x \| Cr^{-1} \| X\varphi_x x)_{XCX11}^1 , \tag{16.5}$$

where the sum takes the values Xx = Aa and Bb. ACA and BCB are degenerate triangles.

The Mulliken approximation of the two-electron integrals can be treated in the same way. Again the approximation must be changed into a correct tensor equation. From the several possibilities we take the simplest one:

$$\left[An_a 1_a^+ m_a, Cn_c 1_c m_c | r_{12}^{-1} | Bn_b 1_b^+ m_b, Dn_d 1_d m_d \right] = \langle An_a l_a m_a | Cn_c l_c m_c \rangle \langle Bn_b 1_b m_b | Dn_d 1_d m_d \rangle$$

$$\cdot (1/4) \sum_{XxYy} \frac{1}{(2l_x+1)(2l_y+1)} \sum_{m_x m_y} \left[Xn_x 1_x^+ m_x, Xn_x 1_x m_x | r_{12}^{-1} | Yn_y 1_y^+ m_y, Yn_y 1_y m_y \right], \tag{16.6}$$

where the sums take the values: Xx = Aa, Cc and Yy = Bb, Dd. This ansatz together with (13.13) leads to the following approximation of the rotational invariants:

$$\left[(An_a 1_a^+, Cn_c 1_c)1 \| (AC^{J_1} BD^{J_2})^J PQ^{J_3} \| (Bn_b 1_b^+, Dn_d 1_d)1 \right]^L = \delta(J_1,1)\delta(J_2,1')$$

$$\cdot \delta(J_3,0)\delta(J,L)(-1)^{l_a+l_b+l_c+l_d+1+1'} \sqrt{1/(21+1)(21+1)} \langle n_a l_a \| AC^l \| n_c l_c \rangle^1$$

$$\cdot \langle n_b l_b \| BD^{l'} \| n_d l_d \rangle^{l'} (1/4) \sum_{XxYy} \sqrt{1/(2l_x+1)(2l_y+1)}$$

$$\cdot \left[(Xn_x 1_x^+, Xn_x 1_x)0 \| (0^\circ 0^\circ)^\circ AB^\circ \| (Yn_y 1_y^+, Yn_y 1_y)0 \right]^\circ \tag{16.7}$$

As in (16.3), most of the invariants (parameters) are put equal to zero.

Without reference to spherical orbitals, the direct approximation of the QRMs is as follows. In (11.2) we put:

$$[Ai\varphi_a a^+m, Ck\varphi_c cp | r_{12}^{-1} | B_j\varphi_b b^+n, Dl\varphi_d dq] = \langle Ai\varphi_a am | Ck\varphi_c cp \rangle \langle Bj\varphi_b bn | Dl\varphi_d dq \rangle \tag{16.8}$$

$$\cdot (1/4) \sum_{Xx i'} \sum_{Yy j'} \frac{1}{dimx \cdot dimy} \cdot \sum_{tu} [Xi'\varphi_x x^+t, Xi'\varphi_x xt | r_{12}^{-1} | Yj'\varphi_y y^+u, Yj'\varphi_y yu]$$

and get the approximative TRMs:

$$[(A\varphi_a a^+, C\varphi_c c)\varepsilon e \| r_{12}^{-1} \| (B\varphi_b b^+, D\varphi_d d)\eta f]_{S\tau g}^{\mu}$$

$$= \frac{\sqrt{Z(-ACS)Z(-BDT)Z(\Delta)}}{dime \cdot dimf} \cdot (A\varphi_a a \| C\varphi_c c)_{S\varepsilon e}^{\varepsilon} \cdot (B\varphi_b b \| D\varphi_d d)_{T\eta f}^{\eta} \cdot PIs^3 \begin{pmatrix} S & S & T \\ \tau & \varepsilon & \eta \\ g & e & f \end{pmatrix}_{\mu} \tag{16.9}$$

$$\cdot (1/4) \sum_{XxYy} \sqrt{1/Z(XYXY)dimx \cdot dimy} \cdot [(X\varphi_x x^+, X\varphi_x x)1 \| r_{12}^{-1} \| (Y\varphi_y y^+, Y\varphi_y y)1]_{XYXY11}^{1}$$

with Xx = Aa, Cc and Yy = Bb, Dd. XYXY are degenerate tetrahedra and thus only two-centre invariants occur.

At this point a comparison with a LCAO-MO parametrization of tetrahedral transition metal complexes (i.e. a MO extension of the ligand field theory) [56] may be of interest. The authors calculate the two-electron integrals of linear combinations of s.-a. MOs:

$$|\gamma \mathring{a} p_a\rangle = c^{\gamma}(\Lambda) \cdot |\Lambda \mathring{a} p_a\rangle + c^{\gamma}(\mathcal{B}) \cdot |\mathcal{B} \mathring{a} p_a\rangle \tag{16.10}$$

where $|\Lambda \mathring{a} p_a\rangle$ and $|\mathcal{B} \mathring{a} p_a\rangle$ are given by (15.3). Set A means the central ion and set B the ligands. They have to discuss the delocalized integrals

$$\langle \gamma_1 \mathring{a} p_a, \gamma_2 \mathring{b} p_b | r_{12}^{-1} | \gamma_3 \mathring{c} p_c, \gamma_4 \mathring{d} p_d \rangle . \tag{16.11}$$

They do not examine the invariants of these integrals according to (11.8/9), but decompose (16.11) into one-electron integrals using the multipole expansion

$$r_{12}^{-1} = \sum_{lm} (4\pi/2l+1)(r_<^l/r_>^{l+1}) \langle \vec{r_1} | lm \rangle \langle lm | \vec{r_2} \rangle . \tag{16.12}$$

Since the delocalized integrals of such radial functions are not calculable, only bilinear combinations of the can be taken into account (tables 11) thus undoing the decomposition.

The bilinear combinations then are parametrized using a manipulated population analysis. Inserting (16.10) and (15.3) into (16.11) yields localized multi-centre integrals, which are subjected to the Mulliken approximation. The result is a long list of separate, numerical coefficients (table 11 again). In order to preserve the point group invariance ad hoc average of the approximate integrals have to be taken.

On the contrary by inserting (16.9) into (11.13), we get a general construction of such coefficients for arbitrary symmetries. Ad hoc averages are not required, not even for orbitals higher than p (cf. [54]).

17. Floating orbitals

A possibility of avoiding the algebra of higher angular momenta is
the exclusive using of s-orbitals floating in space. The invariants
of the preceeding sections depending on the higher quantum numbers thus
are replaced by a multitude of multicentre s-integrals. The task of
the polycentric algebra then is to collect these s-integrals in such
a way, that each independent integral occurs only once in any molecular invariant.

There are two ways of proceeding: The s-orbitals can be distributed
freely in space without regard to the atomic positions and then are
combined to s.-a. MOs. In this case we get high and non-structured multiplicities of the induced symmetry species. On the other hand, the s-
orbitals can be grouped around the atomic positions simulating p-, d-
etc. orbitals. Only from these simulated AOs the s.-a. MOs or VB functions are built up. The attachment of the orbitals to the atomic centres induces a structure into the multiplicity problem, which is accessible to the polyhedral algebra.

To begin with, we simulate orbitals of the symmetry species a by linear combinations of s-orbitals. For this purpose we use one, two or
several equivalent positions $\vec{U_1}$ as required around the centre of symmetry. The s-orbitals are localized at these positions:

$$\langle \vec{r} - \vec{U_1} | \sigma s \rangle = \langle \vec{r} | U1\sigma s \rangle \tag{17.1}$$

σ is a discriminating index, for instance an orbital exponent. According to (5.1) the s.-a. orbitals at the centre of symmetry are given
by:

$$\langle \vec{r} | \text{float } \sigma U \alpha a p_a \rangle = \sum_1 \langle \vec{r} | U1\sigma s \rangle (\vec{U_1} | U \alpha a p_a) \tag{17.2}$$

These compound "AOs" now are shifted to the various atomic positions:

$$\langle \vec{r} | Ai\varphi_a ap_a \rangle = \langle \vec{r} - \vec{A_1} | \text{float } \sigma U \alpha a p_a \rangle , \tag{17.3}$$

where now $\varphi_a = (\text{float } \sigma U a)$.

The invariants of the multi-centre integrals over these shifted orbitals result from the invariants of the integrals over the constituting s-orbitals. As an example, we explicate this for the BRMs. We express the functions (17.3) directly by the twice shifted s-orbitals:

$$\langle \vec{r} | Ai\varphi_a ap_a \rangle = \sum_1 \langle \vec{r} - \vec{A_1} - \vec{U_1} | \sigma s \rangle (\vec{U_1} | U \alpha a p_a)$$

$$= \sum_{lAr} \sqrt{Z(-AAU)} \tau \binom{-AAU}{ril} \langle \vec{r} | Ar\sigma s \rangle (\vec{U_1} | U \alpha a p_a) \tag{17.4}$$

$\vec{A_r}$ are the positions of the s-orbitals and the topological matrix relates them to the atomic positions $\vec{A_i}$.

We now come to the two-centre matrix elements of the orbitals (17.
3/4) where $\varphi_b = (\text{float } \sigma U \beta)$:

$$\langle Ai\varphi_a ap_a | T^C_{p_c} | Bk\varphi_b bp_b \rangle = \sum_{lm} \sum_{\bar{A}r\bar{B}s} \sqrt{Z(-\bar{A}AU)Z(-\bar{B}BU)} \cdot \tau\binom{-\acute{A}AU}{ril} \tau\binom{-\acute{B}BU}{skm}$$

$$\cdot (U\alpha ap_a | \vec{U_1})(\vec{U_m''} | U\beta bp_b) \langle \acute{A}ros | T^C_{p_c} | \acute{B}s\acute{o}s \rangle \qquad (17.5)$$

The integrals on the right are a special case of (4.6):

$$\langle \acute{A}ros | T^C_{p_c} | \acute{B}s\acute{o}s \rangle = \sum_{Vt\gamma} (\acute{A}\acute{o}s \| T^C \| \acute{B}\acute{o}s)_{V\gamma c} \cdot \sqrt{Z(-\bar{A}BV)/dimc} \cdot \tau\binom{-\bar{A}BV}{rst} (\vec{V_t} | V\gamma c p_c) \qquad (17.6)$$

In order to determine the BRMs, we insert this into (17.5) and the result into (4.10):

$$(A\varphi_a a \| T^C \| B\varphi_b b)^{d\delta\eta}_{S\varepsilon e} = dimd \sum_{ikulm\bar{A}r\bar{B}s Vt\gamma} \sum \sum \sqrt{Z(-ABS)Z(-\bar{A}AU)Z(-\bar{B}BU)Z(-\bar{A}BV)/dimc}$$

$$\cdot \binom{a \quad b^+ d}{p_a p_b p_d} \delta \binom{d \quad c^+ e}{p_d p_c p_e} \eta \cdot \tau\binom{-ABS}{iku} \tau\binom{-\acute{A}AU}{ril} \tau\binom{-\bar{B}BU}{skm} \tau\binom{-\acute{A}BV}{rst}$$

$$\cdot (S\varepsilon e p_e | \vec{S_u})(U\alpha ap_a | \vec{U_1})(\vec{U_m''} | U\beta bp_b)(\vec{V_t} | V\gamma c p_c)(\acute{A}\acute{o}s \| T^C \| \acute{B}\acute{o}s)_{V\gamma c} \qquad (17.7)$$

We now inspect the sums for i, k, r, and s of the four topological matrices. Because of the triangular relations (3.1), these sums do not vanish only if $\vec{S_u}+\vec{U_1}=\vec{U_m'}+\vec{V_t}$ or $\vec{U_1}-\vec{U_m'}=\vec{V_t}-\vec{S_u}$. We decompose this quadrangular relation into two triangular relations by introducing the edge vector $\vec{X_y}=\vec{U_1}-\vec{U_m'}$. This edge vector is the distance of two s-orbitals before their shift from the centre to the atomic positions. We therefore get the topological relation:

$$\sum_{ikrs} \sqrt{Z(-ABS)Z(-\bar{A}AU)Z(-\bar{B}BU)Z(-\acute{A}BV)} \cdot \tau\binom{-ABS}{iku} \tau\binom{-\acute{A}AU}{ril} \tau\binom{-\bar{B}BU}{skm} \tau\binom{-\acute{A}BV}{rst} \qquad (17.8)$$

$$= \delta(-\acute{A}AU)\delta(-\bar{B}BU) \sum_{Xy} \sqrt{Z(-UU\acute{X})Z(-VSX)} \cdot \tau\binom{-UU\acute{X}}{lmy} \tau\binom{-VSX}{tuy},$$

where $\delta(-\acute{A}AU) = 1$, if there are triangles of the type $-\acute{A}AU$, and $\delta(-\acute{A}AU) = 0$ otherwise. This relation reminds of the rule of de-Shalit (2.54) with a simple, "topological 9j symbol". We insert (17.8) into (17.7):

$$(A\varphi_a a \| T^C \| B\varphi_b b)^{d\delta\eta}_{S\varepsilon e} = dimd \sum_{ABulmXy Vt\gamma} \sum \delta(-\acute{A}AU)\delta(-\bar{B}BU)\sqrt{Z(-UU\acute{X})Z(-VSX)/dimc}$$

$$\cdot \binom{a \quad b^+ d}{p_a p_b p_d} \delta \binom{d \quad c^+ e}{p_d p_c p_e} \eta \cdot \tau\binom{-UU\acute{X}}{lmy} \tau\binom{-VSX}{tuy}(S\varepsilon e p_e | \vec{S_u})$$

$$\cdot (U\alpha ap_a | \vec{U_1})(\vec{U_m''} | U\beta bp_b)(\vec{V_t} | V\gamma c p_c)(\acute{A}\acute{o}s \| T^C \| \acute{B}\acute{o}s)_{V\gamma c}$$

To the sums for l and m we apply (6.14) and collect the remainder by (6.6). This yields the intended relationship between the BRMs:

$$(A\varphi_a a \| T^C \| B\varphi_b b)^{d\delta\eta}_{S\varepsilon e} = \sum_{ABVX} \sum_{\mu\gamma} \delta(-\acute{A}AU)\delta(-\bar{B}BU)\sqrt{Z(-UU\acute{X})Z(-VSX)/dimc} \qquad (17.9)$$

$$\cdot PIs\begin{pmatrix} -U & U & X \\ \alpha_+ \beta & \mu \\ a^+ b & d | \delta \end{pmatrix} PIs\begin{pmatrix} -V & S & X \\ \gamma & \varepsilon & \mu_+ \\ c & e & d^+ |_\eta \end{pmatrix} \cdot (\acute{A}\acute{o}s \| T^C \| \acute{B}\acute{o}s)_{V\gamma c}$$

The first polyhedral isoscalar characterizes the topology of the un-shifted s-orbitals at the centre, the second one the relation of the atomic distances S to the true distances V of the shifted s-orbitals. The BRMs on the right, of course, depend on the distances V, too. According to (4.10), the BRMs of the s-orbitals result from a simple sum of the integrals:

$$(\acute{A}\sigma s\|T^c\|\acute{B}\acute{\sigma}s)_{V\gamma c} = \sqrt{1/\dim c}\cdot\sum_{\overline{ikp_c}} (V\gamma cp_c|\vec{V}_{1k}) \langle \acute{A}1\sigma s|T^c_{p_c}|\acute{B}k\acute{\sigma}s\rangle$$

$$= \sqrt{1/\dim c}\cdot Z(-\acute{A}\acute{B}V)\sum_{p_c} (V\gamma cp_c|\vec{V}_{1k}) \langle \acute{A}1\sigma s|T^c_{p_c}|\acute{B}k\acute{\sigma}s\rangle, \qquad (17.10)$$

where in the last line the indices of $\vec{V}_{1k}=\vec{A_1}-\vec{B_k}$ are arbitrary.

In the case of scalar operators, follows from (17.9/10):

$$(A\varphi_a a\|T\|B\varphi_b b)^{\delta}_{S\epsilon e} = \sum_{\overline{AB}}\sum_{\overline{VX}}\delta(-\acute{A}AU)\delta(-\acute{B}BU)\cdot\sqrt{Z(-\overline{UUX})Z(-\overline{VSX})Z(-\overline{ABV})}$$

$$\cdot PIs\begin{pmatrix} -U & U & X \\ \alpha_+\beta & \mu \\ a^+b & e & \delta \end{pmatrix} PIs\begin{pmatrix} -V & S & X \\ \epsilon & \mu_+ \\ 1 & e & e^+ \end{pmatrix} \cdot \langle \acute{A}1\sigma s|T|\acute{B}k\acute{\sigma}s\rangle, \qquad (17.11)$$

where $\vec{A_1}-\vec{B_k}$ must be an edge of the type V.

By (17.9) the fact is emphasized, that the existence of molecular multi-centre invariants is not restricted to the linear combinations of spherical atomic orbitals.

18. Overlap- and structural matrices, orthonormalization

The following two sections concern the connection of our new results with the conventional treatment of symmetric molecules. As far as the MO picture[+)] is concerned, we can be brief. Only the VB picture[+)] will show again typical multicentre aspects.

The constructing of many-particle functions and their matrix elements is simplified in both pictures by a preceeding orthonormalization of the one-particle basis. Since the original AOs are not orthogonal, we have to turn over to the delocalized, kanonic MOs or to the localized Löwdin AOs [57]. Apart from series expansions with respect to non-diagonal elements, the diagonalization of the overlap matrix of the original AOs is necessary for both methods. By constructing the s.-a. MOs (5.1) the group theoretical preparations for the diagonalization are done [32]. Because of the orthogonality of the MOs belonging to different symmetry species and components, there remains only the diagonalization of the invariant partial matrices according to:

$$\sqrt{1/\text{dim}\overset{\circ}{a}}\cdot\sum_{\overset{\circ}{\beta}}\langle\overset{\circ}{\kappa}\overset{\circ}{a}\|\overset{\circ}{\beta}\overset{\circ}{a}\rangle\cdot u^{\overset{\circ}{a}}_{\overset{\circ}{\beta}\sigma} = d^{\overset{\circ}{a}}_{\sigma}\cdot u^{\overset{\circ}{a}}_{\overset{\circ}{\kappa}\sigma} \qquad (18.1)$$

The meaning of the quantum numbers is given by (11.6) and the sum runs over all sets of equivalent centres, too. $\overset{\circ}{a}$ represents all occuring symmetry species. $d^{\overset{\circ}{a}}_{\sigma}$ are the eigenvalues and $u^{\overset{\circ}{a}}_{\overset{\circ}{\kappa}\sigma}$ the components of the eigenvectors of the matrices. The reduced matrix elements $\langle\overset{\circ}{\kappa}\overset{\circ}{a}\|\overset{\circ}{\beta}\overset{\circ}{a}\rangle$ result from (15.7), if T is the unit operator.

The kanonical MOs then are given by:

$$|\text{kan.}\sigma\overset{\circ}{a}p_a\rangle = \sum_{\overset{\circ}{\beta}} (d^{\overset{\circ}{a}}_{\sigma})^{-1/2}\cdot u^{\overset{\circ}{a}}_{\overset{\circ}{\beta}\sigma}\cdot|\overset{\circ}{\beta}\overset{\circ}{a}p_a\rangle$$

$$= \sum_{\overset{\circ}{\beta}kn_b} (d^{\overset{\circ}{a}}_{\sigma})^{-1/2}\cdot u^{\overset{\circ}{a}}_{\overset{\circ}{\beta}\sigma}\cdot M(\overset{\circ}{\beta}\overset{\circ}{a}p_a,Bk\beta\overset{''}{b},(\beta b)l_b m_b)\cdot|Bkn_b l_b m_b\rangle \qquad (18.2)$$

Since the eigenvectors belonging to $d^{\overset{\circ}{a}}_{\tau}{=}0$ express the linear dependencies, these cases are omitted in (18.2).

[+)] According to the usage of the quantum theory, unitarily equivalent formulations have to be termed representations or pictures. The Russel-Saunders and the jj coupling, or the strong and the weak field picture of the ligand field theory are unitarily equivalent only if including the full configuration and term interaction. In the same sense, "MO picture" means the construction of MOs with the subsequent full CI calculation, and "VB picture" includes all AO configurations. Omissions and approximations, of course, yield different results within the only basic theory, the quantum mechanics.

Because the eigenvalue problem (18.3) contains invariants only, the eigenvalues are invariants, too. The indirect interrelation between the reduced matrix elements $\langle \mathring{a}\|\mathring{\beta}\mathring{a}\rangle$ and the eigenvalues $d_\sigma^{\mathring{a}}$ can be made some more explicit by the following sum rules:

$$\sum_\sigma (d_\sigma^{\mathring{a}})^n = \mathrm{tr}((S^{\mathring{a}})^n) \quad \text{with} \quad S_{\mathring{A}\mathring{\beta}}^{\mathring{a}} = \langle \mathring{A}\mathring{a}\|\mathring{\beta}\mathring{a}\rangle/\sqrt{\dim \mathring{a}} \quad (18.3)$$

If $S^{\mathring{a}}$ is an m-dimensional matrix, the first m equations of (18.3) fix the eigenvalues. This has been pointed out long ago by Wigner [58]. If we take the sum of (18.3) with respect to $\mathring{a}$, more stringent sum rules result immediately related to the rotational invariants of

$$\langle An_a 1_a m_a | Bn_b 1_b m_b \rangle = \sum_J \sqrt{4\pi}\, \langle n_a 1_a \| AB^J \| n_b 1_b \rangle^J \left({1 \atop m_a}^{J}_{M}\ {1 \atop m_b}^{b} \right) \langle \overline{AB}| \text{ sol } JM \rangle \quad (18.4)$$

a special case of (13.3). The first sums are:

$$\sum_{\sigma \mathring{a}} d_\sigma^{\mathring{a}} = \sum_{An_a 1_a} \langle n_a 1_a \| 0^0 \| n_a 1_a \rangle^0 \cdot Z(A)/\sqrt{2 1_a + 1} \quad (18.5)$$

$$\sum_{\sigma \mathring{a}} (d_\sigma^{\mathring{a}})^2 = \sum_{An_a 1_a}\sum_{Bn_b 1_b}\sum_{TJ} Z(-ABT) \cdot T^{2J} \cdot |\langle n_a 1_a \| T^J \| n_b 1_b \rangle^J|^2 \quad (18.6)$$

Similar rules apply to the eigenvalues of all two-centre matrices.

If we assume a more or less "effective" one-particle Hamiltonian for the electrons of a molecule, similar sum rules result for the one-particle energies $\varepsilon_\sigma^{\mathring{a}}$ belonging to the symmetry species $\mathring{a}$:

$$\sum_\sigma (\varepsilon_\sigma^{\mathring{a}})^n = \mathrm{tr}((\tilde{H}^{\mathring{a}})^n) = \mathrm{tr}((H^{\mathring{a}} S^{\mathring{a}-1})^n) \quad (18.7)$$

The Hamiltonian matrices are given by $\tilde{H}_{\sigma\tau}^{\mathring{a}} = \langle \text{kan.}\sigma \mathring{a}p_a | H | \text{kan.}\tau \mathring{a}p_a \rangle$ and $H_{\mathring{A}\mathring{\beta}}^{\mathring{a}} = \langle \mathring{A}\mathring{a}\|H\|\mathring{\beta}\mathring{a}\rangle/\sqrt{\dim \mathring{a}}$. Because of (18.1) the inverse overlap matrix is:

$$(S^{\mathring{a}-1})_{\mathring{A}\mathring{\beta}} = \sum_\sigma u_{\mathring{A}\sigma}^{\mathring{a}} \cdot (d_\sigma^{\mathring{a}})^{-1} \cdot u_{\mathring{\beta}\sigma}^{\mathring{a}} \quad (18.8)$$

For a preliminary, energetic ordering of the molecular orbitals, a Hückel-like approximation of the two-centre matrix elements $\langle Ain_a 1_a m_a | |H| Bkn_b 1_b m_b \rangle$ may be useful. This is an approximation of the type H = $\alpha I + \beta M$, where M is a structural matrix representing the coordinations, cf [52] section 6.2.2 and [59]. The cited references apply only to s/p_π-orbitals and equal atomic distances and we have to generalize the structural matrix by including different distance vectors and anisotropic orbitals. A characteristic of the Hückel approximation for π-electrons is the independence of the matrix elements from the position of the involved atomic centres. This means that the "topological operator" is regarded as being invariant to translations and rotations. This assumption suffices to fix the form of the structural matrix in accord to the principles of section 13. In analogy to (18.4) follows:

$$\langle An_a 1_a m_a | \text{Top} | Bn_b 1_b m_b \rangle = \sum_J \sqrt{4\pi} \langle n_a 1_a \| AB_J^J \text{Top} \| n_b 1_b \rangle^J \left({1 \atop m_a}^{J}_{M}\ {1 \atop m_b}^{b} \right) \langle \overline{AB}| \text{sol } JM \rangle \quad (18.9)$$

The rotational two-centre invariants $\langle n_a 1_a \| AB^J, \text{Top} \| n_b 1_b \rangle^J$ are the ge-

neralization of the Hückel parameter β. Because of the triangular re-
lation $|l_a-l_b| \leq J \leq l_a+l_b$ and the parity rule $l_a+l_b-J=0$ modulo 2, the
number of the invariants in (18.4 and 9) is equal to the σ-, π-, δ-
etc. bonds of the orbitals at the centres A and B. This interrelation
is made explicit by choosing the distance vector AB parallel to the
z-axis. This produces the standard integrals [60] or standard parameters:

$$\text{Top}_m(n_a l_a n_b l_b, AB) = \langle A n_a l_a m | \text{Top} | B n_b l_b m \rangle_{AB//z} \tag{18.10}$$

$$S_m(n_a l_a n_b l_b, AB) = \langle A n_a l_a m | B n_b l_b m \rangle_{AB//z} ,$$

where in the usual notation $S_0=S_\sigma$, $S_1=S_\pi$, $S_2=S_\delta$ etc. By inverting (18.
4 and 9) the invariants are expressed by the standard integrals:

$$\langle n_a l_a \| AB^J, \text{Top} \| n_b l_b \rangle^J = \sqrt{2J+1} \cdot AB^{-J} \sum_m \left({}_m^{l_a} {}_{0-m}^{J} {}_m^{l_b} \right)(-1)^{l_b-m} \cdot \text{Top}_m(n_a l_a n_b l_b, AB) \tag{18.11}$$

$$\langle n_a l_a \| AB^J \| n_b l_b \rangle^J = \sqrt{2J+1} \cdot AB^{-J} \sum_m \left({}_m^{l_a} {}_{0-m}^{J} {}_m^{l_b} \right)(-1)^{l_b-m} \cdot S_m(n_a l_a n_b l_b, AB)$$

The Hückel-like approximation mentioned above now can be written as:

$$\langle A i n_a l_a m_a | H | B k n_b l_b m_b \rangle = \alpha(n_a l_a) \delta(A,B) \delta(i,k) \delta(n_a,n_b) \delta(l_a,l_b) \delta(m_a,m_b)$$
$$+ \langle A i n_a l_a m_a | \text{Top} | B k n_b l_b m_b \rangle \tag{18.12}$$

Its eigenvalues belonging to the several symmetry species then for instance follow from (18.7). If the number of parameters is still to
large, one might think of the further reduction,

$$\langle n_a l_a \| AB^J, \text{Top} \| n_b l_b \rangle^J = \beta(n_a l_a n_b l_b, AB) \langle n_a l_a \| AB^J \| n_b l_b \rangle^J ,$$

where the overlap invariants may be estimated using the Slater orbitals according to the Slater rules.

The heuristical order of the MOs obtained in this way may serve as
a preparation for the Hartree-Fock-Roothaan approach [61]. Such a preparation is necessary in the case of symmetric systems. The building up
principle requires in this case that we attach a given number of shells
with given occupation numbers to each symmetry species. This is so, because the variation of the orbitals does not alter their symmetry species.

For the purpose of the VB picture on the other hand, orthogonalized
atomic orbitals are of interest. An appropriate basis is that of the
Löwdin orbitals [57]:

$$|\text{Löw.} A i n_a l_a m_a\rangle = \sum_{B k n_b l_b m_b} (S^{-1/2})_{B k n_b l_b m_b, A i n_a l_a m_a} \cdot |B k n_b l_b m_b\rangle \tag{18.13}$$

The problem is the calculation of the root of the inverse overlap matrix. In analogy to (18.8), we first calculate $S^{a-1/2}$, which is the non-
group-theoretical part of the calculation. Then inverting (15.3) by
(15.5), we go back to the angular momentum basis:

$$(S^{-1/2})_{Ain_al_am_a,Bkn_bl_bm_b} = \sum_{\dot{a}\dot{a}\alpha\dot{a}\dot{d}} \sum_{\dot{\beta}\dot{b}\beta\dot{\beta}} \sum_{\dot{\sigma}\dot{a}p_a} M(\ddot{a}\dot{a}p_a, Ai\dot{d}\dot{d}, (\alpha a)1_am_a)$$

$$\cdot u_{\dot{\Lambda}\sigma}^{\dot{a}}(d_\sigma^{\dot{a}})^{-1/2} u_{\dot{\beta}\sigma}^{\dot{a}} \cdot M(\dot{\beta}\dot{a}p_a, Bk\dot{\beta}\dot{b}, (\beta b)1_bm_b) ,$$ (18.14)

where we have to remember $\Lambda = (Ad\dot{d}, n_al_a\alpha a)\dot{\alpha}$. On the other hand, the
bicentric matrix $S^{-1/2}$ is a special case of (4.16) and allows the fol-
lowing representation by two-centre invariants:

$$(S^{-1/2})_{Ain_al_am_a,Bkn_bl_bm_b} = \sum_{\dot{a}\alpha\dot{\beta}\dot{b}} \sum_{\dot{\sigma}\varepsilon e} S^{-1/2}(An_al_a\dot{a}\dot{a}\| Bn_bl_b\dot{\beta}\dot{b})_{S\varepsilon e}^\delta$$

$$\cdot \langle 1_am_a|1_a\dot{a}\dot{a}p_a \rangle \langle 1_b\dot{\beta}\dot{b}p_b|1_bm_b \rangle (\substack{\dot{a}+\dot{b}'\ e\\ p_ap_bp_e}\substack{\delta\\})\sqrt{1/dime}(\vec{S}_{ik}|S\varepsilon ep_e) ,$$ (18.15)

where the BRMs result from the eigenvalues and eigenvectors of (18.1):

$$S^{-1/2}(An_al_a\alpha a\| Bn_bl_b\beta b)_{S\varepsilon e}^\delta = \sqrt{Z(-ABS)dime} \cdot \sum_{\dot{a}\alpha\beta\dot{b}} \sum_{\eta\dot{a}\dot{a}\sigma} dim\dot{a}\{bb'\dot{a}^+\dot{\beta}\}\{ab^+e\delta\}$$

$$\cdot \{\substack{\dot{a}^+\dot{b}'\ e\\ b\ \ a\ \dot{a}^+}\}_{\alpha\beta\delta\eta} \cdot PIs\begin{pmatrix}-A & B & S\\ \alpha' & \beta' & \varepsilon\\ a' & b^+ & e^+\end{pmatrix}_\eta \cdot u_{\dot{\Lambda}\sigma}^{\dot{a}}(d_\sigma^{\dot{a}})^{-1/2}u_{\dot{\beta}\sigma}^{\dot{a}}$$ (18.16)

In essence this is an inversion of (5.13).

19. Many-electron systems

19.1. Summary of the occupation operator technique

A concise description of symmetric many-electron systems is achieved by using the occupation operators, as has been shown by Judd in the case of atomic spectroscopy [62]. But the conventional approach to this technique is quite fussy. There is no need of the detour via a special one-particle basis, determinantal functions, the occupation number representation and so on. This applies especially to symmetric systems, the states of which in general are no determinants at all. We therefore define the creation and annihilation operators with respect to an arbitrary orbital $|\varphi\rangle$ directly by their effect on an arbitrary N-particle function $|N\Phi\rangle$. Such a state is only restricted by the fermion property:

$$T(i,k)\langle r_1 \ldots r_N | N\Phi\rangle = (-1) \cdot \langle r_1 \ldots r_N | N\Phi\rangle , \qquad (19.1)$$

where $T(i,k)$ means the transposition of r_i and r_k:

$$T(i,k)\langle r_1 \ldots r_{i-1} r_i r_{i+1} \cdots r_{k-1} r_k r_{k+1} \cdots r_N | N\Phi\rangle$$
$$= \langle r_1 \ldots r_{i-1} r_k r_{i+1} \cdots r_{k-1} r_i r_{k+1} \cdots r_N | N\Phi\rangle \qquad (19.2)$$

The operators $a^+(\varphi)$ and $a(\varphi)$ adding and annihilating an electron in the orbital $|\varphi\rangle$ now are defined by

$$a^+(\varphi)|N\Phi\rangle = |N+1\Phi\varphi\rangle \qquad (19.3)$$
$$a(\varphi)|N\Phi\rangle = |N-1\Phi\varphi\rangle , \qquad (19.4)$$

where the (N+1)- and (N-1)-particle functions are given by

$$\langle r_1 \ldots r_{N+1} | N+1\Phi\varphi\rangle = (1/\sqrt{N+1})\sum_i (-1)^{N+1-i} T(N+1,i)\langle r_{N+1}|\varphi\rangle\langle r_1 \ldots r_N|N\Phi\rangle \qquad (19.5)$$
$$\langle r_1 \ldots r_{N-1} | N-1\Phi\varphi\rangle = \sqrt{N}\int \langle \varphi | r_N\rangle\langle r_1 \ldots r_N|N\Phi\rangle d^3 r_N \qquad (19.6)$$

The consistency of these definitions requires the proof, that $a^+(\varphi)$ and $a(\varphi)$ are adjoint operators, i.e.

$$\langle N+1\Psi | a^+(\varphi)|N\Phi\rangle = \langle N\Phi | a(\varphi)|N+1\Psi\rangle^* \qquad (19.7)$$

with arbitrary $|N+1\Psi\rangle$ and $|N\Phi\rangle$. (Assuming two determinants differing just by the one orbital $|\varphi\rangle$ makes the proof at least very incomplete.) The proof is as follows:

$$\langle N+1\Psi | a^+(\varphi)|N\Phi\rangle = (1/\sqrt{N+1})\sum_i \int \ldots \int \langle N+1 | r_1 \ldots r_{N+1}\rangle$$
$$\cdot (-1)^{N+1-i} T(N+1,i)\langle r_{N+1}|\varphi\rangle\langle r_1 \ldots r_N|N\Phi\rangle d^3 r_1 \ldots d^3 r_{N+1}$$

Since the notation of the integration variables is arbitrary, we interchange some of them:

$$= (1/\sqrt{N+1})\sum_i \int \ldots \int (-1)^{N+1-i} \left[T(N+1,i)\langle N+1\Psi | r_1 \ldots r_{N+1}\rangle \right]$$
$$\cdot \langle r_{N+1} | \varphi\rangle \langle r_1 \ldots r_N | N\Phi\rangle d^3 r_1 \ldots d^3 r_{N+1}$$

Because of (19.1):

$$= \sqrt{N+1} \int \ldots \int \langle N+1\, \psi | r_1 \ldots r_{N+1} \rangle \langle r_{N+1} | \varphi \rangle \langle r_1 \ldots r_N | N\phi \rangle d^3 r_1 \ldots d^3 r_{N+1}$$

and with (19.6):

$$= \int \ldots \int \langle a(\varphi)(N+1\,\psi) | r_1 \ldots r_N \rangle \langle r_1 \ldots r_N | N\phi \rangle d^3 r_1 \ldots d^3 r_N = \langle a(\varphi)(N+1\,\psi) | N\phi \rangle ,$$

which completes the proof.

From (19.1-6) further follow the well-known commutation relations:

$$[a^+(\varphi), a^+(\eta)]_+ = 0 \qquad (19.8)$$

$$[a(\varphi), a(\eta)]_+ = 0 \qquad (19.9)$$

$$[a^+(\varphi), a(\eta)]_+ = \langle \eta | \varphi \rangle \qquad (19.10)$$

The last equation indicates the complications arising from the use of non-orthogonal one-particle bases. For this reason, we have set up the kanonical and the Löwdin bases in the preceeding section and presuppose an orthogonal basis $\langle i | k \rangle = \delta_{ik}$ in the following. With respect to such a basis, we have the well-known operator representations

$$T = \sum_{ik} \langle i | t | k \rangle a^+(i) a(k) \qquad (19.11)$$

for a one-particle operator and

$$G = (1/2) \sum_{iklm} \langle ik | g | lm \rangle \, a^+(i) a^+(k) a(l) a(m) \qquad (19.12)$$

for a two-particle operator. The essential point in the proof of (19.11) is the expansion

$$T = \sum_{i=1}^{N} t_i = \sum_{ikl} \langle r_i' | k \rangle \langle k | t | l \rangle \langle l | r_i \rangle$$

in combination with (19.6/7). The proof of (19.12) requires the similar expansion of G. The matrix elements of arbitrary N-particle states then are:

$$\langle N\phi | T | N\psi \rangle = \sum_{ik} \langle i | t | k \rangle \langle N\phi | a^+(i) a(k) | N\psi \rangle \qquad (19.13)$$

and

$$\langle N\phi | G | N\psi \rangle = \frac{1}{2} \sum_{iklm} \langle ik | g | lm \rangle \langle N\phi | a^+(i) a^+(k) a(l) a(m) | N\psi \rangle , \qquad (19.14)$$

i.e. weighted sums of the one- and two-particle matrix elements. The weight factors are called one- and two-particle density matrices. If the many-particle functions are built up by a systematic, for instance recursive, calculus, the density matrix elements are pure geometric and combinatorial coefficients, cf.eq.(19.23). This scheme now has to be transferred to the symmetrized MO and VB picture.

19.2. Group theory and occupation operators

For the purpose of the MO picture, we use for instance the kanonical orbitals (18.2) as a starting point. These delocalized orbitals have the same transformation properties as the orbitals of one-centre expansions and approximations. We therefore can take over the aufbau principle of the strong field coupling. But in contrast to the atomic spectroscopy, there is no universal energetic order of the orbitals.

This is demonstrated by the example discussed in [52], page 444. The chemical intuition is supported by the topological considerations of the preceeding section.

The following is known from the literature [5, 8, 9, 16, 17]. Combining group theory and occupation operators we follow the cited ref. [62]. A brief indication concerning point groups is given in [6]. The main interest of the following repetition of [16, 17] is to get a model for the VB picture. As in these references, the quantum numbers are optionally single numbers for G' in $\gamma\gamma$-coupling or double indices for $G \times SU(2)$ in ΓS-coupling (cf. section 2.9.). Thus there is no need to mention the spin explicitly. The main point in the sense of this treaty is that (19.13 and 14) again turn into interrelations between reduced many-particle and reduced one- and two-particle matrix elements mediated again by geometric and combinatorial factors.

The occupation operators of a s.-a. kanonical (spin) orbital (18.2) $|\text{kan.}\mathring{\sigma}\mathring{a}p_a\rangle$ are $a^+(\text{kan.}\mathring{\sigma}\mathring{a}p_a)$ and $a(\text{kan.}\mathring{\sigma}\mathring{a}p_a)$. In the following, we do not explicitly mention the index kan., because it does not affect the group theory. The operators are (double) tensor operators transforming in accordance with the representations $\mathring{a}$ and $\mathring{a}^+$ (note the dagger). Since the proof given in [6] does not apply to point groups, we show the transformation property of $a(\mathring{\sigma}\mathring{a}p_a)$. In analogy to (2.6), the unitary operator representation of $g \in G$ is defined by

$$\langle r_1 \ldots r_N | U(g) | N\Phi \rangle = \langle g^{-1}r_1 \ldots g^{-1}r_N | N\Phi \rangle \tag{19.15}$$

We thus have:
$$U(g)a(\mathring{\sigma}\mathring{a}p_a)|N\Phi\rangle = U(g)|N-1\overline{\Phi}\mathring{\sigma}\mathring{a}p_a\rangle$$
$$= \int \ldots \int |r_1 \ldots r_{N-1}\rangle\langle r_1 \ldots r_{N-1}|U(g)|N-1\overline{\Phi}\mathring{\sigma}\mathring{a}p_a\rangle d^3r_1 \ldots d^3r_{N-1}$$

Because of (19.15):
$$= \int \ldots \int |r_1 \ldots r_{N-1}\rangle\langle g^{-1}r_1 \ldots g^{-1}r_{N-1}|N-1\overline{\Phi}\mathring{\sigma}\mathring{a}p_a\rangle d^3r_1 \ldots d^3r_{N-1}$$

and with (19.6):
$$= \int \ldots \int |r_1 \ldots r_{N-1}\rangle\langle\mathring{\sigma}\mathring{a}p_a|g^{-1}r_N\rangle\langle g^{-1}r_1 \ldots g^{-1}r_N|N\Phi\rangle d^3r_1 \ldots d^3r_N \cdot \sqrt{N}$$

With (19.15) again:
$$= \int \ldots \int |r_1 \ldots r_{N-1}\rangle\langle\mathring{\sigma}\mathring{a}p_a|g^{-1}r_N\rangle\langle r_1 \ldots r_N|U(g)|N\Phi\rangle d^3r_1 \ldots d^3r_N \cdot \sqrt{N}$$

Further with (2.17):
$$= \sum_{q_a} D^{\mathring{a}}_{q_a p_a}(g)^* \int \ldots \int |r_1 \ldots r_{N-1}\rangle\langle\mathring{\sigma}\mathring{a}q_a|r_N\rangle\langle r_1 \ldots r_N|U(g)|N\Phi\rangle d^3r_1 \ldots d^3r_N \cdot \sqrt{N}$$

and (19.6) again:
$$= \sum_{q_a} D^{\mathring{a}}_{q_a p_a}(g)^* \cdot a(\mathring{\sigma}\mathring{a}p_a)U(g)|N\Phi\rangle$$

Since $|N\Phi\rangle$ is arbitrary, we conclude:

$$U(g)a(\sigma \mathring{a} p_a) = \sum_{q_a} D^{\mathring{a}}_{q_a p_a}(g)^* \cdot a(\sigma \mathring{a} q_a) U(g) \, ,$$

or with (2.5):

$$U(g)a(\sigma \mathring{a} p_a)U(g)^+ = \sum_{q_a} D^{(\mathring{a}^+)}_{q_a p_a}(g) \cdot a(\sigma \mathring{a} q_a) \, , \qquad (19.16)$$

which has to be shown.

The commutators (19.8-10) now appear as:

$$\left[a^+(\sigma \mathring{a} p_a), a^+(\tau \mathring{b} p_b) \right]_+ = \left[a(\sigma \mathring{a} p_a), a(\tau \mathring{b} p_b) \right]_+ = 0$$

$$\left[a^+(\sigma \mathring{a} p_a), a(\tau \mathring{b} p_b) \right]_+ = \delta(\sigma, \tau)\delta(\mathring{a}, \mathring{b})\delta(p_a, p_b) \qquad (19.17)$$

The $\delta(\sigma, \tau)$ does not result from group theory, but from the choice of (18.2).

The many-particle functions are built up in priciple recursively:

$$|N{+}1\varepsilon e p_e\rangle = \{e\}\sqrt{\text{dime}} \sum_{p_a p_d p_e} \binom{\mathring{a}^+ \mathring{d}^+ e}{p_a p_d p_e}^{\eta} \cdot a^+(\sigma \mathring{a} p_a) |N\delta d p_d\rangle \qquad (19.18)$$

with $\varepsilon = (\delta d \sigma \mathring{a} \eta)$. In order to select an orthonormal set from the functions defined in (19.18), the godparent scheme in connection with Löwdins orthogonalization has been proposed in [16]. As an alternative one may think of the seniority or quasi-spin formalism [63]. Further it is advisable to use (19.18) only within each shell and to combine the (open or closed) shells afterwards, cf. [16] section 6. But the details are not relevant in the present context.

If we apply the WET to the creation operators, we get:

$$\langle N\varepsilon e p_e | a^+(\sigma \mathring{a} p_a) |N{-}1\delta d p_d\rangle = \sum_{\eta} \langle N\varepsilon e \| a^+(\sigma \mathring{a}) \| N{-}1\delta d\rangle_{\eta} \binom{e^+ \mathring{a}\ d}{p_e p_a p_d}^{\eta} \qquad (19.19)$$

Except for a factor, the RMEs are the coefficients of fractional parentage (CFP):

$$\langle N\varepsilon e \| a^+(\sigma \mathring{a}) \| N{-}1\delta d\rangle_{\eta} = (-1)^N \sqrt{N \cdot \text{dime}} \langle N\varepsilon e \{ | N{-}1\delta d, \sigma \mathring{a} \rangle_{\eta} \qquad (19.20)$$

This relation shows that the concept of fractional parentage is not restricted to certain recursive schemes. It makes sense in much more general circustances. We give an extreme example: $|N\varepsilon e p_e\rangle$ may be a strong field coupling function, $|N{-}1\delta d p_d\rangle$ a weak field coupling function both built up from GTOs, and $a^+(\sigma \mathring{a} p_a)$ the creator of a STO. Because of (19.5), the CFP can be calculated, if the functions are given explicitly. If on the other hand a recursive construction according to (19.18) is used, the CFPs can be calculated recursively without explicit knowledge of the many-particle functions.

The analogue of (19.11) then is:

$$T^c_{p_c} = \sum_{\sigma_1 \mathring{a}_1 p_1} \sum \langle \sigma_1 \mathring{a}_1 p_1 | t^c_{p_c} | \sigma_2 \mathring{a}_2 p_2 \rangle a^+(\sigma_1 \mathring{a}_1 p_1) a(\sigma_2 \mathring{a}_2 p_2)$$

$$= \sum_{\sigma_1 \mathring{a}_1 p_1 \varepsilon} \sum \langle \sigma_1 \mathring{a}_1 \| t^c \| \sigma_2 \mathring{a}_2 \rangle_{\varepsilon} \binom{\mathring{a}^+_1 c\ \mathring{a}_2}{p_1 p_c p_2}^{\varepsilon} a^+(\sigma_1 \mathring{a}_1 p_1) a(\sigma_2 \mathring{a}_2 p_2) \qquad (19.21)$$

From (19.20/21) results the general interrelation between the reduced many-particle and the reduced one-particle matrix elements:

$$\langle N\delta_1 d_1 \| T^c \| N\delta_2 d_2 \rangle_\eta = \sum_{\sigma_1 \mathring{a}_1 \epsilon} \langle \sigma_1 \mathring{a}_1 \| t^c \| \sigma_2 \mathring{a}_2 \rangle_\epsilon \cdot g^{N\eta\epsilon}_{\delta_1 d_1, \delta_2 d_2}(\sigma_1 \mathring{a}_1, \sigma_2 \mathring{a}_2, c) \quad (19.22)$$

with the weight factor:

$$g^{N\eta\epsilon}_{\delta_1 d_1, \delta_2 d_2}(\sigma_1 \mathring{a}_1, \sigma_2 \mathring{a}_2, c) = N\sqrt{dim d_1 dim d_2} \sum_{\varphi f \sigma \tau} \langle N-1\varphi f, \sigma_1 \mathring{a}_1 \| N\delta_1 d_1 \rangle_\sigma$$

$$\cdot \langle N-1\varphi f, \sigma_2 \mathring{a}_2 \| N\delta_2 d_2 \rangle_\tau \{f \mathring{a}_1 d_1^+ \sigma\} \{\mathring{a}_1^+ c \mathring{a}_2 \epsilon\} \{\mathring{a}_2\} \left\{ \begin{matrix} c & d & d_1^+ \\ f^+ & \mathring{a}_2 & \mathring{a}_1 \end{matrix} \right\}_{\epsilon \tau \sigma \eta} \quad (19.23)$$

This again is a relation of the principal type (1.2) with the geometrical factor (19.23). We have termed it g.. keeping to the notation of Griffith [5], section 7.2.

Similar relations result for the two-particle operators. (19.12)
reads now:

$$G = (1/2) \sum_{\sigma_1 \mathring{a}_1 p_1} \sum \langle \sigma_1 \mathring{a}_1 p_1, \sigma_2 \mathring{a}_2 p_2 | g | \sigma_3 \mathring{a}_3 p_3, \sigma_4 \mathring{a}_4 p_4 \rangle$$

$$a^+(\sigma_1 \mathring{a}_1 p_1) a^+(\sigma_2 \mathring{a}_2 p_2) a(\sigma_3 \mathring{a}_3 p_3) a(\sigma_4 \mathring{a}_4 p_4)$$

$$= (1/2) \sum_{\sigma_1 \mathring{a}_1 p_1 \gamma\mu c} \sum \langle (\sigma_1 \mathring{a}_1, \sigma_2 \mathring{a}_2) \gamma c \| g \| (\sigma_3 \mathring{a}_3, \sigma_4 \mathring{a}_4) \mu c \rangle \quad (19.24)$$

$$\cdot A^+((\sigma_1 \mathring{a}_1, \sigma_2 \mathring{a}_2) \gamma c p_c) A((\sigma_3 \mathring{a}_3, \sigma_4 \mathring{a}_4) \mu c p_c)$$

with the pair operators (s.-a. geminal creation operators):

$$A^+((\sigma_1 \mathring{a}_1, \sigma_2 \mathring{a}_2) \gamma c p_c) = \sqrt{dim c} \sum_{p_1 p_2 p_c} \begin{pmatrix} \mathring{a}_1^+ \mathring{a}_2^+ c \\ p_1 p_2 p_c \end{pmatrix}^\gamma \cdot a^+(\sigma_1 \mathring{a}_1 p_1) a^+(\sigma_2 \mathring{a}_2 p_2) \quad (19.25)$$

From these operators follow the two-particle CFP:

$$\langle N\epsilon e \| A^+((\sigma_1 \mathring{a}_1, \sigma_2 \mathring{a}_2) \gamma c) \| N-2\delta d \rangle_\eta$$

$$= \sqrt{N(N-1) dim e} \langle N\epsilon e \{ \| N-2\delta d, (\sigma_1 \mathring{a}_1, \sigma_2 \mathring{a}_2) \gamma c \rangle_\eta \quad (19.26)$$

And the matrix elements of the two-particle operator finally are:

$$\langle N\delta_1 dp | G | N\delta_2 dp \rangle = \frac{N(N-1)}{2} \sum_{\sigma_1 \mathring{a}_1 \gamma\mu c} \sum \langle (\sigma_1 \mathring{a}_1, \sigma_2 \mathring{a}_2) \gamma c \| g \| (\sigma_3 \mathring{a}_3, \sigma_4 \mathring{a}_4) \mu c \rangle$$

$$\cdot \sum_{\varphi f \eta} \langle N-2\varphi f, (\sigma_1 \mathring{a}_1, \sigma_2 \mathring{a}_2) \gamma c \| N\delta_1 d \rangle_\eta \langle N-2\varphi f, (\sigma_3 \mathring{a}_3, \sigma_4 \mathring{a}_4) \mu c \| N\delta_2 d \rangle_\eta \quad (19.27)$$

again a relation of type (1.2).

As a consequence of the delocalization of the kanonical MOs, the
results, shortly summarized here, have no specific polycentric character. This character again appears, if we calculate the density matrix of a molecular state with respect to a localized AO basis, in order to read the charge and bond orders with respect to this basis
[64]. We only discuss the density matrices of totally symmetric
states or the average density of degenerate states. Taking the average is equivalent to picking out the totally symmetric part of the
density matrix [65]:

$$\Gamma_{A i \varphi_a a p_a, B k \varphi_b b p_b} = dim d^{-1} \sum_{p_d} \langle N\alpha d p_d | a^+(A i \varphi_a a p_a) a(B k \varphi_b b p_b) | N\delta d p_d \rangle \quad (19.28)$$

Because of (19.19/20), this is essentially a bilinear sum of CFP in
the AO basis. As a typical two-centre matrix, (19.28) falls within the
scope of (4.13) and has the following representation:

$$\Gamma_{Ai\varphi_a ap_a, Bk\varphi_b bp_b} = \sum_{\delta\varepsilon e} \Gamma(A\varphi_a a\| B\varphi_b b)^\delta_{S\varepsilon e} \binom{a^+ b\ e}{p_a p_b p_e}^\delta \sqrt{1/\text{dime}}(\vec{S}_{ik}|S\varepsilon e p_e) \quad (19.29)$$

The BRMs with S$\neq$0 are the symmetry-invariant representation of the
bond orders classified according to the bond edges, those with S=0
the charge orders. This is the solution of the problem posed in [64],
page 60, i.e. the problem to define the bond orders invariantly to
symmetry operations in the presence of several bonding electrons.

20. Sketch of the VB picture

In contrast to the MO picture, the VB picture naturally contributes further stimulating aspects to the polycentric algebra. After the initial success by Heitler and London the VB picture [66] has not been popular because of the multitude of configurations. But the calculations, focussed on the atoms and compounds of the first period, have led to a one-sided view. Among the transition-metal and lanthanide compounds treated in the first approximation by the weak field coupling, there are obvious candidates for VB calculations and moreover of its atoms-in-molecules version proposed by Moffitt [67]. In this connection, we again point to the discussion in [52], page 444. In view of the expenditure on CI calculations the VB may become competitive again.

Avoiding orthogonality problems, we start from the Löwdin basis (18. 13) and adapt it to the symmetry group:

$$|\text{Löw.Ai}\varphi_a ap_a\rangle = \sum |\text{Löw.Ain}_a l_a m_a\rangle \langle l_a m_a | l_a \alpha ap_a\rangle \qquad (20.1)$$

with $\varphi_a = (n_a l_a \alpha)$. The commutation relations of the pertinent occupation operators are:

$$[a^+(\text{Löw.Ai}\varphi_a ap_a), a^+(\text{Löw.Bk}\varphi_b bp_b)]_+ = [a(\text{Löw.Ai}\varphi_a ap_a), a(\text{Löw.Bk}\varphi_b bp_b)]_+ = 0$$

$$[a^+(\text{Löw.Ai}\varphi_a ap_a), a(\text{Löw.Bk}\varphi_b bp_b)]_+ = \delta(A,B)\delta(i,k)\delta(\varphi_a,\varphi_b)\delta(a,b)\delta(p_a,p_b)$$

$$\qquad (20.2)$$

where in detail $\delta(\varphi_a,\varphi_b) = \delta(n_a,n_b)\delta(l_a,l_b)\delta(\alpha,\beta)$.

After the model of (19.18), one generates the VB functions by repeated application of the creation operators. Each added (spin) orbital introduces a new centre into the many-particle function, so that every VB function is associated with a more or less asymmetric polyhedron. The vertices, some of which may coincide again, are valued differently by the AOs.

Against the first sight, this building up does not lead to an incalculable multitude of different polyhedra because of two reasons. By (3.25), (7.8), and (10.2) we already have become aquainted with the equivalence of polyhedra and vectors inducing the same representation of the symmetry group. There is only a limited number of such representations, which are determined by the vectors a) in general position, b) on equivalent plains of reflection, and c) on equivalent axes of rotation. Because of this equivalence, the polyhedra fall into classes inducing equivalent representations. Fully identical representations are achieved only, if the numbering of the polyhedra is coordinated.

On the other hand, the number of different polyhedra is limited by the Pauli principle. If all orbitals $|\text{Ai}\varphi_a ap_a\rangle$ with fixed A, φ_a, and a are doubly occupied, the result is a full supershell. We use this term with respect to reducible representations. In general a super-

shell contains several ordinary shells belonging to irreducible representations. The polyhedron associated with the supershell has 2·dima·
$Z(A)$ vertices, 2·dima of which coincide at each centre. Therefore it
is totally symmetric, and there are no higher polyhedra within a supershell.

In the following, P^n represents a set of equivalent polyhedra with
n vertices and P_k^n the k-th polyhedron of this set. The symmetry-adap-
tation of a scalar polyhedral function $(P_k^n|F)$ is analogous to (3.17):

$$(P^n \varepsilon ep_e|F) = \sum_k (P^n \varepsilon ep_e|P_k^n)(P_k^n|F) , \qquad (20.3)$$

where the SALC coefficients $(P^n \varepsilon ep_e|P_k^n)$ are determined via the asso-
ciated vectors $\overrightarrow{P_k^n} = \sum_{A_i} \mu_{Ai} \cdot \overrightarrow{A_i}$ with $\overrightarrow{A_i} \in P_k^n$.

In parallel to (19.18), we now generate the many-particle VB states:

$$|Q_1^{n+1} \delta dp_a\rangle = \{d\} \sqrt{Z(Q^{n+1})} dimd \sum_k \binom{a^+c^+d}{P_aP_cP_d}^\varepsilon \cdot \hat{\tau}\binom{A \quad P^nQ^{n+1}}{i \quad k \quad 1} \cdot a^+(L\delta w \cdot A\varphi_a ap_a)|P_k^n \gamma cp_c\rangle \qquad (20.4)$$

with $\delta = (A\varphi_a a, P^n \gamma c, \varepsilon)$. The quantum numbers δ and γ recursively notify
the antecedents of the state. $\hat{\tau}$ is the generalization of our topolo-
gical matrices and does not vanish only, if Q_1^{n+1} comes from P_k^n by the
addition of the new vertex at A_1:

$$\hat{\tau}\binom{A \quad P^nQ^{n+1}}{i \quad k \quad 1} = \begin{cases} 1/\sqrt{Z(Q^{n+1})}, & \text{if } P_k^n \dotplus \overrightarrow{A_i} = Q_1^{n+1} \\ 0 & \text{otherwise} \end{cases} \qquad (20.5)$$

Obviously there are orthogonality relations like:

$$\sum_{ik} \hat{\tau}\binom{A \quad P^nQ^{n+1}}{i \quad k \quad 1} \cdot \hat{\tau}\binom{A \quad P^nR^{n+1}}{i \quad k \quad m} = \delta(Q,R)\delta(1,m)/Z(Q^{n+1}) \qquad (20.6)$$

Apart from the antisymmetrization, the formation of the geminals (9.3)
is a simple example of (20.4). The states generated in this way are
in general non-orthogonal and moreover often linearly dependent. Of
cours, the diagonalization of the overlap matrix of these state may
serve for the orthonormalization again. But because of the high dimen-
sion of the supershells, other methods, like the quasi-spin or seniority
formalism, will be more economic. The functions (20.4) transform in
accord to the reducible product representation $d \times \sigma^{P^n}$.

If T_p^d is a translation invariant or a s.-a. operator at the centre
of symmetry, the matrix elements $\langle Q_k^n \varepsilon ep_e|T_p^d|R_1^n \acute{e}ep_e'\rangle$ are associated
with a polyhedron of 2n vertices composed of the polyhedra Q_k^n and R_1^n.
We express this composition by another topological relation:

$$\hat{\tau}\binom{Q^nR^nP^{2n}}{k \quad 1 \quad m} = \begin{cases} 1/\sqrt{Z(P^{2n})}, & \text{if } Q_k^n \dotplus R_1^n = P_m^{2n} \\ 0 & \text{otherwise} \end{cases} \qquad (20.7)$$

Having defined this, we can state the following theorem:

$$\langle Q_k^n \varepsilon e p_e | T_{P_d}^d | R_1^n \acute{e} \acute{e} p_e' \rangle = \sum_{f\eta\varphi Pm\pi p} (Q^n \varepsilon e \| T^d \| R^n \acute{e} \acute{e}) \frac{f\eta\varphi}{P^{2n}\pi p} \cdot \binom{e^+ e' \ f}{P_e P_e' P_f}^\eta \binom{f^+ d \ p^+ \varphi}{P_f P_d P_p}$$

$$\cdot \sqrt{Z(P^{2n})} \hat{\tau} \binom{Q^n R^n P^{2n}}{k \ 1 \ m} \cdot (P_m^{2n} | P^{2n} \pi p p_p) \qquad (20.8)$$

This theorem, including the proof, agrees totally with (4.6). The invariants therefore are termed polyhedral, reduced matrix elements (PRM).

The polyhedral VB functions (20.4) now are combined to SALCs. This is quite analogous to (5.1):

$$|Q^n \varphi f p_f\rangle = \sum_{k p_e} \hat{K}(\eta f p_f, Q^n k \delta d, e p_e) \cdot |Q_k^n \varepsilon e p_e\rangle \qquad (20.9)$$

with $\varphi = (\varepsilon e, \delta d, \eta)$, where ε is a compound index corresponding with (20.4). The generalized SALC coefficient is:

$$\hat{K}(\eta f p_f, Q^n k \delta d, e p_e) = \{f\} \sqrt{\dim f} \sum \binom{d^+ e^+ f}{P_d P_e P_f}^\eta \cdot (Q_k^n | Q^n \delta d p_d) \qquad (20.10)$$

Apart from (5.1), we already know another example of (20.9), namely the formation of the s.-a. geminals (9.4).

Again we apply the WET to the matrix elements of the states (20.9)

$$\langle Q^n \varphi f p_f | T_{P_c}^c | R^n \gamma g p_g \rangle = \sum_\varepsilon \langle Q^n \varphi f \| T^c \| R^n \gamma g \rangle_\varepsilon \binom{f^+ c \ g}{P_f P_c P_g}^\varepsilon \quad , \qquad (20.11)$$

the RMEs of which are related to the PRMs quite in parallel to the relations (5.5 or 8). This relation requires the definition of a polyhedral isoscalar generalizing (6.6):

$$\widehat{PIs} \begin{pmatrix} Q^n R^n P^{2n} \\ \delta \ g \ \pi \\ d \ r \ p \end{pmatrix} = \sum_{\varepsilon} \sum_{k l m} \hat{\tau} \binom{Q^n R^n P^{2n}}{k \ 1 \ m} (Q_k^n | Q^n \delta d p_d)(R_1^n | R^n \text{or} p_r)(P_m^{2n} | P^{2n} \pi p p_p) \binom{d \ r \ p}{P_d P_r P_p}^{\varepsilon^*} \qquad (20.12)$$

Then follows the theorem:

$$\langle Q^n \varphi f \| T^c \| R^n \gamma g \rangle_{\varepsilon'} = \sum_z \widehat{GEO}_1(x, y, z) \cdot (Q^n \varepsilon e \| T^c \| R^n \acute{e} \acute{e}) \frac{h\sigma\tau}{P^{2n}\pi p} \qquad (20.13)$$

with the generalized geometrical factor

$$\widehat{GEO}_1(x, y, z) = \{f^+ \text{gce}\} \sqrt{Z(P^{2n})} \dim f \cdot \dim g \cdot \sum_\alpha \widehat{PIs} \begin{pmatrix} Q^n R^n P^{2n} \\ \delta \ \acute{\delta} \ \pi \\ d^+ d' \ p \end{pmatrix}_\alpha \cdot \begin{cases} f^+ g \ c \\ d \ d'^+ p^+ \\ e \ \acute{e}^+ h \end{cases}_\alpha \begin{matrix} \varepsilon'' \\ \alpha \\ \sigma \end{matrix} \qquad (20.14)$$

and the compound indices $\varphi = (\varepsilon e, \delta d, \eta)$, $\gamma = (\acute{e} \acute{d}, \acute{\delta} \acute{d}, \theta)$, $y = (\delta d \eta f, \acute{\delta} d \theta g, \varepsilon')$, $x = (Q^n e, R^n e', c)$, and $z = (h \sigma \tau P^{2n} \pi p)$.

Besides (5.5), another example of (20.13) is hidden in (11.20), since the geometrical factor GEO_3 is composed of three factors of the general type GEO_1. From the general point of view the polyhedral isoscalar of the third kind is nothing but

$$PIs^3 \begin{vmatrix} \mathcal{J} \ S \ T \\ . \ . \ . \\ . \ . \ . \end{vmatrix} = \widehat{PIs} \begin{vmatrix} T^4 S^2 T^2 \\ . \ . \ . \\ . \ . \ . \end{vmatrix}. \qquad (20.15)$$

The 6j symbol appears in (11.20) instead of a 9j symbol as a result of (2.52/53), because the operator is a scalar.

One-particle operators have the representation

$$T_{P_c}^c = \sum_{A1Bk} \sum_{\varphi_x x p_x} \langle L\ddot{o}w.A1\varphi_a a p_a | t_{P_c}^c | L\ddot{o}w.Bk\varphi_b b p_b \rangle$$
$$\cdot a^+(L\ddot{o}w.A1\varphi_a a p_a) a(L\ddot{o}w.Bk\varphi_b b p_b) \qquad (20.16)$$

and in order to trace back the PRMs to the one-particle matrix elements and their invariants, we must examine the polyhedral matrix elements of the occupation operators, i.e. $\langle Q_1^{n+1}\delta dp_d|a^+(L\ddot{o}w.Ai\varphi_a ap_a)|$ $|P_k^n \varepsilon ep_e\rangle$. These matrix elements are zero, if $P_k^n + \vec{A_i} = R_m^{n+1} \neq Q_1^{n+1}$, for the VB functions referring to R_m^{n+1} and Q_1^{n+1} are mutually orthogonal, since they differ by one Löwdin orbital. Hence the matrix elements behave like the topological matrices (20.5) and allow the factorization:

$$\langle Q_1^{n+1}\delta dp_d|a^+(L\ddot{o}w.Ai\varphi_a ap_a)|P_k^n \varepsilon ep_e\rangle = C\cdot\hat{\tau}\left(\begin{matrix} Q^{n+1} & A & P^n \\ 1 & i & k \end{matrix}\right)$$

The proportionality factor C transforms according to the direct product $d^+ \times a \times e$, so that the WET yields a further factorization:

$$\langle Q_1^{n+1}\delta dp_d|a^+(L\ddot{o}w.Ai\varphi_a ap_a)|P_k^n \varepsilon ep_e\rangle \tag{20.17}$$
$$= \sum_\mu \langle Q^{n+1}\delta d||a^+(A\varphi_a a)||P^n \varepsilon e\rangle_\mu \left(\begin{matrix} d^+ a & e \\ p_d p_a p_e \end{matrix}\right)^\mu \cdot \hat{\tau}\left(\begin{matrix} Q^{n+1} & A & P^n \\ 1 & i & k \end{matrix}\right)$$

At this point, one might argue that the matrix elements transform according to the sixfold product $d^+ \times a \times e \times \sigma^{Q^{n+1}} \times \sigma^A \times \sigma^{P^n}$ and should have a more complex coupling structure than (20.17). But since the couplings of the polyhedra and of the orbital representations have been kept strictly apart in (20.4), the matrix behaves like a double tensor of $i^+ \times a \times e$ on one hand and of $\sigma^{Q^{n+1}} \times \sigma^A \times \sigma^{P^n}$ on the other.

In analogy to (19.20), the polyhedral CFP are defined as follows:

$$\langle Q^{n+1}\delta d||a^+(A\varphi_a a)||P^n \varepsilon e\rangle_\mu = (-1)^{n+1}\sqrt{(n+1)Z(Q^{n+1})\mathrm{dimd}}\langle Q^{n+1}\delta d\ \{|P^n \varepsilon e, A\varphi_a a\rangle_\mu \tag{20.18}$$

Having defined this, we can determine the PRMs of (20.8) by the BRMs of the Löwdin orbitals, a relation that replaces (19.22) in the VB case. For this purpose, we solve (20.8) for the PRMs:

$$(Q^n \varepsilon e||T^c||R^n \varepsilon' e')_{p\pi p}^{f\eta\varphi} = \sqrt{Z(P^{2n})}\cdot\mathrm{dimf}\cdot\sum_{klm}(P^{2n}\pi pp_p|P_m^{2n})\cdot\hat{\tau}\left(\begin{matrix} Q^n R^n P^{2n} \\ k & l & m \end{matrix}\right)$$
$$\cdot\left(\begin{matrix} f^+c & p_p^+\varphi^* \\ p_f p_c p_p \end{matrix}\right)\left(\begin{matrix} e^+e' & f \\ p_e p_{e'} p_p \end{matrix}\right)\eta^*\langle Q^n \varepsilon ep_e|T_{p_c}^c|R^n \varepsilon' e'p_{e'}\rangle \tag{20.19}$$

For the matrix elements on the right side, we substitute (20.16) and (4.6) for the one-particle matrix elements in (20.16). This yields:

$$(Q^n \varepsilon e||T^c||R^n \varepsilon' e')_{p\pi p}^{f\eta\varphi} = \sum_y GEO_g(x,y)\cdot(L\ddot{o}w.A\varphi_a a||t^c||L\ddot{o}w.B\varphi_b b)_{S\sigma s}^{g\gamma\alpha} \tag{20.20}$$

with $x=(Q^n \varepsilon e, R^n \varepsilon' e', f\eta\varphi, P^{2n}\pi p)$ and $y=(A\varphi_a a, B\varphi_b b, S\sigma s, g\gamma\alpha)$. For the present, the geometrical factor is given by:

$$EO_g(x,y) = \sqrt{Z(P^{2n})Z(-ABS)}\mathrm{dimf}\cdot\sum_{klm}\sum_{ijt}(P^{2n}\pi pp_p|P_m^{2n})\hat{\tau}\left(\begin{matrix} Q^n R^n P^{2n} \\ k & l & m \end{matrix}\right)\left(\begin{matrix} f^+c & p_p^+\varphi^* \\ p_f p_c p_p \end{matrix}\right.$$
$$\cdot\left(\begin{matrix} e^+e' & f \\ p_e p_{e'} p_f \end{matrix}\right)\eta^*\left(\begin{matrix} a^+b & g \\ p_a p_b p_g \end{matrix}\right)\gamma\left(\begin{matrix} g^+c & s_s^+ \\ p_g p_c p_s \end{matrix}\right)\alpha\cdot\tau\left(\begin{matrix} -ABS \\ i & j & t \end{matrix}\right)(\vec{S_t}|S\sigma sp_s)$$
$$\cdot\langle Q_k^n \varepsilon ep_e|a^+(L\ddot{o}w.Ai\varphi_a ap_a)a(L\ddot{o}w.Bj\varphi_b bp_b)|R_1^n \varepsilon' e'p_{e'}\rangle$$

between the two occupation operators we insert a complete function system, i.e. $\sum_{Tr}\sum_{\delta dp_d}|T_r^{n-1}\delta dp_d\rangle\langle T_r^{n-1}\delta dp_d|$, express the resulting matrix elements by (20.17) and collect the 3jm symbols in a 6j symbol:

$$GEO_9(x,y) = \delta(f,g)\delta(s,p)\delta(\alpha,\varphi)\text{dims}^{-1}\sqrt{Z(P^{2n})Z(-ABS)}\{ba^+f\gamma\}\sum_{\delta d\mu\mu}\{e'^+bd\mu'\}$$

$$\cdot\begin{Bmatrix}e & a^+d^+\\b & e'\ f^+\end{Bmatrix}_{\eta\gamma\mu'\mu}\cdot\sum_{klmi}\sum_{jtTr}(P^2\pi sp_s|P_m^2)(\vec{S}_t|So sp_s)\hat{\tau}(\begin{smallmatrix}Q^n R^n P^{2n}\\k\ l\ m\end{smallmatrix})\tau(\begin{smallmatrix}-ABS\\i j t\end{smallmatrix})$$

$$\cdot\hat{\tau}(\begin{smallmatrix}Q^n A\ T^{n-1}\\k\ i\ r\end{smallmatrix})\hat{\tau}(\begin{smallmatrix}R^n B\ T^{n-1}\\l\ j\ r\end{smallmatrix})\langle Q^n\varepsilon e\| a^+(A\varphi_a a)\| T^{n-1}\delta d\rangle_\mu\langle R^n\varepsilon'e'\| a^+(B\varphi_b b)\| T^{n-1}\delta d\rangle_{\mu'}$$

From this expression, we isolate the following topological invariant, which is remotely similar to a 6j symbol:

$$PW\begin{Bmatrix}Q & R & P^\pi/S^{\sigma s}\\B & A & T\end{Bmatrix} = \text{dims}^{-1}\sum_{klmi}\sum_{jtr}\hat{\tau}(\begin{smallmatrix}Q^n R^n P^{2n}\\k\ l\ m\end{smallmatrix})\tau(\begin{smallmatrix}-ABS\\i j t\end{smallmatrix})\hat{\tau}(\begin{smallmatrix}Q^n A\ T^{n-1}\\k\ i\ r\end{smallmatrix})\hat{\tau}(\begin{smallmatrix}R^n B\ T^{n-1}\\l\ j\ r\end{smallmatrix})$$

$$\cdot(\vec{S}_t|So sp_s)(P^{2n}\pi sp_s|P_m^{2n}) \quad (20.21)$$

It relates the edges S associated with the one-particle BRMs to the polyhedra P^{2n} belonging to the many-particle PRMs. Speaking more precisely, it indicates, which edges S interrelate the centres in Q^n with those in R^n and if the representations σ^S and $\sigma^{P^{2n}}$ have in common the irreducible representation s.

Introducing the polyhedral CFP by (20.18) yields the final expression of the geometrical factor:

$$GEO_9(x,y) = \delta(f,g)\delta(s,p)\delta(\alpha,\varphi)\cdot n\cdot\sqrt{Z(P^{2n})Z(-ABS)Z(Q^n)Z(R^n)\text{dime}\cdot\text{dime}'}$$

$$\cdot\{ba^+f\gamma\}\sum_{T\delta d\mu\mu'}\{e'^+bd\mu\}'\begin{Bmatrix}e & a^+d^+\\b & e'\ f^+\end{Bmatrix}_{\eta\gamma\mu\mu'}\cdot PW\begin{Bmatrix}Q & R & P^\pi/S^{\sigma s}\\B & A & T\end{Bmatrix} \quad (20.22)$$

$$\cdot\langle Q^n\varepsilon e\| T^{n-1}\delta d, A\varphi_a a\rangle_\mu\langle R^n\varepsilon'e'\| T^{n-1}\delta d, B\varphi_b b\rangle_{\mu'}$$

By inserting (20.20) into (20.13), we can directly link the RMEs of the s.-a. VB functions to the one-particle BRMs:

$$\langle Q^n\varphi f\| T^c\| R^n\varphi'f\rangle_{\varepsilon'} = \sum_{y_2} GEO_{10}(x_1,y_1,y_2)\cdot(\text{L\"ow}.A\varphi_a a\| t^c\| \text{L\"ow}.B\varphi_b b)_{So s}^{g\gamma\alpha} \quad (20.23)$$

with $\varphi=(\varepsilon e,\delta d,\eta)$, $\varphi'=(\varepsilon'e',\delta d',\eta')$, $x_1=(Q^n\ e,R^n\ e',c)$, $y_1=(\delta d\eta f,\delta d'\eta'f',\varepsilon')$, and $y_2=(A\varphi_a a,B\varphi_b b,So s,g\gamma\alpha)$ and the compound factor:

$$GEO_{10}(x_1,y_1,y_2) = \sum_{h\sigma\tau P^{2n}\pi p}\widetilde{GEO}_1(x_1,y_1,z_1)\cdot GEO_9(z_2,y_2), \quad (20.24)$$

where $z_1=(h\sigma\tau P^{2n}\pi p)$ and $z_2=(Q^n\varepsilon e,R^n\varepsilon'e',h\sigma\tau,P^{2n}\pi p)$.

Of course, a similar analysis involving the polyhedral two-particle CFP can be made for the two-particle interaction operators.

21. Prospect of crystals

As mentioned in the introduction, the symmetry analysis lined out, so far, can be transferred to crystals, i.e. to space groups. To begin with, we confine the discussion to the formation of SALCs, as far as this can be achieved without using projective representations [68].

21.1. The irreducible representations of space groups

We summarize the concepts needed to set up the irreducible representations of space groups. In general we keep to the book of Streitwolf [37].

The space group is marked by G, the related point group by $G_0 \sim G/T$, where T is the translation group. An element $\{\alpha|\vec{a}\}$ of a space group is the operation defined by a translation $\vec{a} \in T$ and a rotation $\alpha \in G_0$:

$$\{\alpha|\vec{a}\}\vec{r} = \alpha\vec{r} + \vec{a} \tag{21.1}$$

From this definition follows $\{\beta|\vec{b}\}\{\alpha|\vec{a}\} = \{\beta\alpha|\beta\vec{a}+\vec{b}\}$ and $\{\alpha|\vec{a}\}^{-1} = \{\alpha^{-1}|-\alpha^{-1}\vec{a}\}$. The unitary operator representing $\{\alpha|\vec{a}\} \in G$ in the function space is in accordance with (4.2) defined by:

$$\langle\vec{r}|U(\{\alpha|\vec{a}\})|\varphi\rangle = \langle\{\alpha|\vec{a}\}^{-1}\vec{r}|\varphi\rangle = \langle\alpha^{-1}\vec{r}-\alpha^{-1}\vec{a}|\varphi\rangle = \langle\alpha^{-1}(\vec{r}-\vec{a})|\varphi\rangle \tag{21.2}$$

The irreducible representations of the space groups are characterized by wave vectors within the first Brillouin zone. To each wave vector $\vec{k}$ belongs a subgroup $G_{0\vec{k}} \subset G_0$, the elements of which leave $\vec{k}$ invariant or transform it into an equivalent wave vector:

$$G_{0\vec{k}} = \{\beta \text{ with } \beta \in G_0 \text{ and } \beta\vec{k} = \vec{k} + \vec{K}\}, \tag{21.3}$$

where $\vec{K}$ means a lattice vector of the reciprocal lattice. This point group is also termed the little cogroup of $\vec{k}$ [68]. The irreducible, projective representations b of these cogroups enter the ordinary, irreducible representations of the space groups. We designate the irreducible, projective representations of the little cogroups by

$$D^{\vec{k}b}_{p_b q_b}(\beta) \quad \text{with } \beta \in G_{0\vec{k}}, \tag{21.4}$$

where $\vec{k}$ indicates the pertinent wave vector. Because this concept of projective representations is significant only for some wave vectors at the surface of the Brillouin zone of non-symmorphic space groups, it is entirely avoided by Streitwolf [37]. But it allows the general and concise formulation of all irreducible representations of all space groups (21.10) below, cf. [68].

In order to select the elements of $G_{0\vec{k}}$ from those of G_0, we define the symbol
$$\Delta^{\vec{k}}(\alpha) = \begin{cases} 1 & \text{if } \alpha \in G_{0\vec{k}}, \text{ i.e. } \alpha\vec{k}=\vec{k}+\vec{K} \\ 0 & \text{if } \alpha \notin G_{0\vec{k}}, \text{ i.e. } \alpha\vec{k}\neq\vec{k}+\vec{K} \end{cases} \tag{21.5}$$

The coset decomposition of G_0 with respect to $G_{0\vec{k}}$ is:

$$G_0 = \sum_i \alpha_i G_{o\vec{k}} \tag{21.6}$$

with the coset representatives α_i.

By applying these representatives to $\vec{k}$, one generates a set of wave vectors $\vec{k}_j$ (within the first Brillouin zone), the so called star of $\vec{k}$, which is designated by ^+k:

$$^+k = \{\vec{k}_j \text{ with } \vec{k}_j = \alpha_j \vec{k} + \vec{K}\} \tag{21.7}$$

We choose $\alpha_1 = \epsilon$, i.e. $\vec{k}_1 = \vec{k}$. The vectors $\vec{k}_j \epsilon^+k$ are termed the prongs of the star.

We now can generalize (21.5) to all $\alpha \epsilon G_0$ by defining

$$\Delta_{ij}^{\vec{k}}(\alpha) = \Delta^{\vec{k}}(\alpha_i^{-1}\alpha\alpha_j), \tag{21.8}$$

or because of (21.5 and 7):

$$\Delta_{ij}^{\vec{k}}(\alpha) = \begin{cases} 1 & \text{if } \alpha\vec{k}_j = \vec{k}_i + \vec{K} \\ 0 & \text{if } \alpha\vec{k}_j \neq \vec{k}_i + \vec{K} \end{cases} \tag{21.9}$$

Now all necessary notations are collected to write down all irreducible representations of the space groups. These are determined by a star ^+k and an irreducible (projective) representation b of $G_{o\vec{k}}$ with $\vec{k}\epsilon^+k$. The components of the representations are determined likewise by a double index, i.e. by the prong j ($\vec{k}_j$ respectively) and by the component p_b of $b(G_{o\vec{k}})$. Using (2.4), the representation matrices $D^{(^+kb)}(\{\alpha|\vec{a}\})$ then are given by:

$$D_{ip_b, jr_b}^{(^+kb)}(\{\alpha|\vec{a}\}) = \Delta_{ij}^{\vec{k}}(\alpha) D_{p_b r_b}^{\vec{k}b}(\alpha_i^{-1}\alpha\alpha_j) \cdot \exp(-i\vec{k}_i \cdot \vec{a}) \tag{21.10}$$

The associated bases are denoted $|\beta(^+kb)ip_b\rangle$, where β is a discriminating index.

At this point, we must insert a marginal note concerning the phase in eq. (21.10). This phase is sometimes chosen more complicate as in [69], eq.(4.19) and [70b], eq.(I.4). The difference results from two gauge transformations, A) of the representations of $G_{o\vec{k}}$: $D^{\vec{k}b}(\beta) = \exp(-i\vec{k}\cdot\vec{v}_\beta)\cdot D^{\vec{k}b}(\beta)$ and B) of the bases: $|\beta(^+kb)ip_b\rangle = \exp(-i\vec{k}_i\cdot\vec{v}_{\alpha_i})\cdot|\beta(^+kb)ip_b\rangle$, where $\vec{v}_\gamma$ denotes the non-primitive translation belonging to the rotation or reflection γ. We have three reasons for our choice of the phase: A) for internal points of the Brillouin zone our matrices $D^{\vec{k}b}(\beta)$ are genuine vector representations and not only projectively-equivalent to vector representations, cf.eq.(4.25) of [69]. B) In the following relations concerning the bases (21.13 and 14) no phase factors occur. C) Eq. (21.10) is plainly simpler.

The transformation property of the bases with respect to the operation (21.2) is now given by:

$$U(\{\alpha|\vec{a}\})|\beta(^+kb)jr_b\rangle = \sum_{ip_b} D_{ip_b, jr_b}^{(^+kb)}(\{\alpha|\vec{a}\}) \cdot |\beta(^+kb)ip_b\rangle \tag{21.11}$$

The SALCs or tight-binding functions intended have to comply with this relation. Since we have derived the theory of SALC coefficients for ordinary vector representations only, we must exclude the few cases requiring really projective representations.

21.2. Symmetry-adapted functions for symmorphic space groups

We start with a construction, which is restricted to symmorphic space groups but immediately resumes the molecular symmetry-adaption; for in the the symmorphic case the Wigner-Seitz unit cell can be treated like a molecule and the point-group adaption can be simply combined with the well-known Bloch sums [37], eq.(6.14).

The relations (21.10 and 11) suggest that the atomic orbitals must be adapted to the pertinent little cogroup. The coupling according to (5.2) then has to be done with respect to the same group. The adaptation thus depends on the point of the Brillouin zone.

In order not to operate with SALC coefficients, 3jm symbols, and s.-a. atomic orbitals of several point groups simultaneously, we proceed as follows. All atomic orbitals are classified according to the irreducible representations of G_0 and in the frame of a universal, fixed coordinate system. Then the s.-a. Wigner-Seitz-cell orbitals are formed using the SALC coefficients belonging to G_0. And only in conclusion, the SALCs are subduced to $G_{0\vec{k}}$ if necessary. If $\vec{A_i}$ are the position vectors of the equivalent atoms within the Wigner-Seitz cell, the s.-a. cell orbitals according to (5.1/2) are given by:

$$\langle \vec{r} | (A\varepsilon e, \varphi_a a) \gamma c p_c \rangle = \sum_{ip_a} K^0(\gamma c p_c, Ai\varepsilon e, ap_a) \cdot \langle \vec{r} - \vec{A_i} | \varphi_a a p_a \rangle$$

with

$$K^0(\gamma c p_c, Ai\varepsilon e, ap_a) = \{c\}\sqrt{dimc} \sum_{p_e} \binom{e^+a^+c}{p_e p_a p_c}^{\gamma}(\vec{A_i} | A\varepsilon e p_e)$$

For $\vec{k}=0$, G_0 is the little cogroup and the s.-a. tight-binding functions are simple Bloch sums for the lattice vectors $\vec{R}$:

$$\langle \vec{r} | \delta(^+0c)0p_c \rangle = \sum_{Rip_a} K^0(\gamma c p_c, Ai\varepsilon e, ap_a) \langle \vec{r} - \vec{A_i} - \vec{R} | \varphi_a a p_a \rangle \qquad (21.12)$$

with $\delta = (\varphi_a a A\varepsilon e \gamma)$.

In the other cases, the s.-a. Wigner-Seitz-cell orbitals have to be subduced to the little cogroup $G_{0\vec{k_j}}$ belonging to the relative prong $\vec{k_j}$. All little cogroups $G_{0\vec{k_j}}$ belonging to the same star are isomorphic to $G_{0\vec{k}}$:

$$G_{0\vec{k_j}} = \alpha_j^{-1} G_{0\vec{k}} \alpha_j$$

This means for the representation matrices:

$$D^{\vec{k_j}b}_{p_b q_b}(\beta_j) = D^{\vec{k}b}_{p_b q_b}(\alpha_j^{-1}\beta\alpha_j) \quad \text{with } \beta_j \epsilon G_{0\vec{k_j}}, \ \beta \epsilon G_{0\vec{k}}$$

The coordinate axes are different for each $\vec{k}_j$ and the bases are related by

$$U(\alpha_j)|\vec{k}bp_b\rangle = |\vec{k}_j bp_b\rangle \tag{21.13}$$

Hence it follows for the adaption coefficients of the group chains $G_o \supset G_{o\vec{k}_j}$ and $G_o \supset G_{o\vec{k}}$:

$$\langle cp_c|\vec{k}_j\beta bp_b\rangle = \sum_{q_c} D^{oc}_{p_c q_c}(\alpha_j)\langle cq_c|\vec{k}\beta bp_b\rangle, \tag{21.14}$$

where c is a representation of G_o with the matrices $D^{oc}(\alpha)$.

We no state that the s.-a. tight-binding functions are given by

$$\langle\vec{r}|\delta(^+kb)jp_b\rangle \tag{21.15}$$

$$= \sum_{RI}\sum_{p_a p_c}\langle cp_c|\vec{k}_j\beta bp_b\rangle K^o(\gamma cp_c, A i\epsilon e, ap_a)\cdot \exp(i\vec{k}_j\cdot\vec{R})\cdot\langle\vec{r}-\vec{A}_i-\vec{R}|\varphi_a ap_a\rangle$$

with $\delta=(a\varphi_a A\epsilon e\gamma c\beta)$.

To prove this, we have to demonstrate the property (21.11/10). With (21.2), we have at first:

$$\langle\vec{r}|U(\{\alpha|\vec{a}\})|\delta(^+kb)jp_b\rangle \tag{21.16}$$

$$=\sum_{RI}\sum_{p_a p_c}\exp(i\vec{k}_j\cdot\vec{R})\langle cp_c|\vec{k}_j\beta bp_b\rangle K^o(\gamma cp_c, A i\epsilon e, ap_a)\langle\alpha^{-1}(\vec{r}-\vec{a}-\alpha\vec{A}_i-\alpha\vec{R})|\varphi_a ap_a\rangle$$

Because of (4.2) and (3.3), it follows

$$\langle\alpha^{-1}(\vec{r}-\vec{a}-\alpha\vec{A}_i-\alpha\vec{R})|\varphi_a ap_a\rangle= \sum_{kq_a} D^{oa}_{q_a p_a}(\alpha)\sigma^A_{ki}(\alpha)\langle\vec{r}-\vec{a}-\vec{A}_k-\alpha\vec{R}|\varphi_a aq_a\rangle \tag{21.17}$$

and further with (3.5) and (2.22):

$$\sum_{Ip_a} K^o(\gamma cp_c, A i\epsilon e, ap_a)\cdot D^{oa}_{q_a p_a}(\alpha)\sigma^A_{ki}(\alpha) = \sum_{q_c} K^o(\gamma cq_c, A k\epsilon e, aq_a)\cdot D^{oc}_{q_c p_c}(\alpha) \tag{21.18}$$

Substituting (21.17/18) into (21.16) yields the intermediate result

$$\langle\vec{r}|U(\{\alpha|\vec{a}\})|\delta(^+kb)jp_b\rangle = \sum_{Rkp_c}\sum_{q_a q_c} \exp(i\vec{k}_j\cdot\vec{R})\langle cp_c|\vec{k}_j\beta bp_b\rangle D^{oc}_{q_c p_c}(\alpha)$$

$$\cdot K^o(\gamma cq_c, A k\epsilon e, aq_a)\langle\vec{r}-\vec{a}-\vec{A}_k-\alpha\vec{R}|\varphi_a aq_a\rangle. \tag{21.19}$$

This requires the calculation of the sum

$$\sum_{p_c} D^{oc}_{q_c p_c}(\alpha)\langle cp_c|\vec{k}_j\beta bp_b\rangle$$

$$= \sum_{Ip_c r_c s_c} D^{oc}_{q_c r_c}(\alpha_i)D^{oc}_{r_c s_c}(\alpha_i^{-1}\alpha\alpha_j)D^{oc}_{s_c p_c}(\alpha_j^{-1})\langle cp_c|\vec{k}_j\beta bp_b\rangle\Delta^{\vec{k}}_{ij}(\alpha) ,$$

where $\alpha_i^{-1}\alpha\alpha_j \epsilon G_{o\vec{k}}$ because of (21.9). Using (21.14), we get further:

$$= \sum_{Ir_c s_c} D^{oc}_{q_c r_c}(\alpha_i)D^{oc}_{r_c s_c}(\alpha_i^{-1}\alpha\alpha_j)\langle cs_c|\vec{k}\beta bp_b\rangle\Delta^k_{ij}(\alpha)$$

$$= \sum_{Ir_c q_b} D^{oc}_{q_c r_c}(\alpha_i)\langle cr_c|\vec{k}\beta bq_b\rangle D^{kb}_{q_b p_b}(\alpha_i^{-1}\alpha\alpha_j)\Delta^k_{ij}(\alpha)$$

and finally with (21.14) again:

$$\sum_{p_c} D^{oc}_{q_c p_c}(\alpha)\langle cp_c|\vec{k}_j\beta bp_b\rangle = \sum_{Iq_b} \langle cq_c|\vec{k}_j\beta bp_b\rangle D^{kb}_{q_b p_b}(\alpha_i^{-1}\alpha\alpha_j)\Delta^k_{ij}(\alpha) \tag{21.20}$$

We have to insert this result into (21.19). In addition, we substitute $\vec{a}+\alpha\vec{R}=\vec{R}'$ or $\vec{R}=\alpha^{-1}(\vec{R}'-\vec{a})$. Replacing the sum for $\vec{R}$ by a sum for $\vec{R}'$, we get:

$$\langle\vec{r}|U(\{\alpha|\vec{a}\})|\delta(^+kb)jp_b\rangle = \sum_{\overline{Rk1q_x}} \exp(i\alpha\vec{k}_j\cdot(\vec{R}-\vec{a}))\cdot D^{\vec{kb}}_{q_bp_b}(\alpha_i^{-1}\alpha\alpha_j)\cdot\Delta^{\vec{k}}_{ij}(\alpha)$$

$$\cdot\langle cq_c|\vec{k}_1\beta bq_b\rangle K^0(\gamma cq_c,Ak\epsilon e,aq_a)\langle\vec{r}-\vec{A}_k-\vec{R}|\varphi_a aq_a\rangle$$

Because of (21.7), $\alpha_i^{-1}\alpha\alpha_j \in G_{o\vec{k}}$, and (21.3), we have:

$$\alpha\vec{k}_j=\alpha_i(\alpha_i^{-1}\alpha\alpha_j)\alpha_j^{-1}\vec{k}_j=\alpha_i(\alpha_i^{-1}\alpha\alpha_j)\vec{k}+\vec{k}_1=\alpha_i\vec{k}+\vec{k}_2+\vec{k}_1=\vec{k}_1+\vec{k}_3+\vec{k}_2+\vec{k}_1=\vec{k}_1+\vec{k}$$

Since $\vec{R}-\vec{a}$ is a lattice vector in the symmorphic case, this yields:

$$\exp(i\alpha\vec{k}_j\cdot(\vec{R}-\vec{a})) = \exp(i\vec{k}_i(\vec{R}-\vec{a}))$$

and we can sum up with (21.15):

$$\langle\vec{r}|U(\{\alpha|\vec{a}\})|\delta(^+kb)jp_b\rangle=\sum_{\overline{1q_b}}\Delta^{\vec{k}}_{ij}(\alpha)D^{\vec{kb}}_{q_bp_b}(\alpha_i^{-1}\alpha\alpha_j)\exp(-i\vec{k}_i\cdot\vec{a})\langle\vec{r}|\delta(^+kb)iq_b\rangle$$

This proves (21.11/10).

21.3. Construction includung non-symmorphic groups

The Wigner-Seitz cell of non-symmophic groups is not invariant to G_o and in general not to $G_{o\vec{k}}$. Therefore we can no more rely directly upon the SALC coefficients of the point groups in the configuration space as in (21.12).

We propose another method likewise applicable to symmorphic and non-symmorphic groups. To this end we remember section 12. Accoding to (12.16), the SALC coefficients could be determined by substituting the atomic positions into s.-a. functions. If we set aside the ortho-normalization in a first step, every s.-a. may serve for this purpose.

Such s.-a. functions are the symmetrized plain waves according to [37], section 6.2. For internal points of the Brillouin zone, we can set up these symmetrized plain waves again by the help of the SALC coefficients using them now in the reciprocal lattice. We again start with k=0 in order to keep the calculation in G_o descending to G_{ok} in the end.

We choose a set of equivalent lattice vectors in general position. This choice avoids invariance groups $G_{o\vec{K}}$ and their cosets and yields a maximal number of functions. Starting with a lattice vector $\vec{K}$ in general position, we can number the members of the equivalent set by the elements $\beta \in G_o$: $\vec{K}_\beta=\beta\vec{K}$. We now claim that the following SALC of plain waves belongs to the irreducible representation (^+0e) of the space group:

$$\langle\vec{r}|W(^+0e)0p_e\rangle = \sum_{\beta\in G_o} (\vec{K}_\beta|K\epsilon ep_e)\cdot\exp(-\vec{K}_\beta\cdot\vec{v}_\beta)\cdot\exp(i\vec{K}_\beta\cdot\vec{r}) \qquad (21.21)$$

The proof of (21.21) is contained in that of (21.23).

The symmetrized plain waves of the general case $\vec{k}\neq0$ are:

$$\langle\vec{r}|W\epsilon\beta(^+kb)jp_b\rangle=\sum_{\beta p_e}(\vec{K}_\beta|K\epsilon ep_e)\langle ep_e|\vec{k}_j\beta bp_b\rangle\exp(-i\vec{k}_\beta\cdot\vec{v}_\beta)\exp[i(\vec{K}_\beta+\vec{k}_j)\cdot\vec{r}] \qquad (21.22)$$

But in order apply the coupling coefficients of G_o, only we use (21.21) to form the s.-a. tight-binding functions for the special case $\vec{k}=0$:

$$\langle\vec{r}|TB\delta(^+0c)0p_o\rangle=\sum_{Rt\beta'p_a}K^o(\gamma cp_c,K\beta'\epsilon e,ap_a)\exp\left[i\vec{K}_{\beta'}\cdot(\vec{R}+\vec{A}_t-\vec{v}_{\beta'})\right]\langle\vec{r}-\vec{R}-\vec{A}_t|\varphi_a ap_a\rangle$$

with $\delta=(\varphi_a aAK\epsilon e\gamma)$. Descending now to $G_{o\vec{k}_j}$ we generate the s.-a. tight-binding functions for the general case:

$$\langle\vec{r}|TB\eta(^+kb)jp_b\rangle = \sum_{Rt\beta'}\sum_{p_a p_c}K^o(\gamma cp_c,K\beta'\epsilon e,ap_a)\langle cp_c|\vec{k}_j\beta bp_b\rangle \tag{21.23}$$
$$\cdot\exp\left[i\vec{K}_{\beta'}\cdot(\vec{R}+\vec{A}_t-\vec{v}_\beta)\right]\exp\left[i\vec{K}_j\cdot(\vec{R}+\vec{A}_t)\right]\langle\vec{r}-\vec{R}-\vec{A}_t|\varphi_a ap_a\rangle$$

with $\eta=(\varphi_a aAK\epsilon e\gamma c\beta)$.

In our final proof, we show that this formula comprises all tight-binding functions of internal points of the first Brillouin zone. At first we have with (21.2):

$$\langle\vec{r}|U(\{\alpha|\vec{a}\})|TB\eta(^+kb)jp_b\rangle = \sum_{Rt\beta'}\sum_{p_a p_c}K^o(\gamma cp_c,K\beta'\epsilon e,ap_a)\langle cp_c|\vec{k}_j\beta bp_b\rangle$$
$$\cdot\exp\left[i\vec{K}_{\beta'}\cdot(\vec{R}+\vec{A}_t-\vec{v}_{\beta'})\right]\exp\left[i\vec{K}_j\cdot(\vec{R}+\vec{A}_t)\right]\langle\alpha^{-1}(\vec{r}-\vec{a}-\alpha\vec{R}-\alpha\vec{A}_t)|\varphi_a ap_a\rangle$$

Substituting $\vec{a}+\alpha\vec{R}+\alpha\vec{A}_t=\vec{R}+\vec{A}_u$ or $\vec{R}+\vec{A}_t=\alpha^{-1}(\vec{R}+\vec{A}_u-\vec{a})$ and using (4.2) yields:

$$= \sum_{Ru\beta'}\sum_{p_a p_c q_a}K^o(\gamma cp_c,K\beta'\epsilon e,ap_a)\langle cp_c|\vec{k}_j\beta bp_b\rangle\cdot\exp\left[i\alpha\vec{K}_{\beta'}\cdot(\vec{R}+\vec{A}_u-\vec{a}-\alpha\vec{v}_{\beta'})\right]$$
$$\cdot\exp\left[i\alpha\vec{K}_j\cdot(\vec{R}+\vec{A}_u-\vec{a})\right]\cdot D^{oa}_{q_a p_a}(\alpha)\langle\vec{r}-\vec{R}-\vec{A}_u|\varphi_a aq_a\rangle$$

Because of $\{\alpha|\vec{a}\}\{\beta|\vec{v}_\beta\} = \{\alpha\beta'|\vec{a}+\alpha\vec{v}_{\beta'}\} = \{\alpha\beta|\vec{v}_{\alpha\beta}+\vec{R}'\}$, there is the relation

$$\exp\left[i\alpha\vec{K}_{\beta'}\cdot(\vec{R}+\vec{A}_u-\vec{a}-\alpha\vec{v}_{\beta'})\right] = \exp\left[i\alpha\vec{K}_{\beta'}\cdot(\vec{R}+\vec{A}_u-\vec{v}_{\alpha\beta'})\right]$$
$$= \sum_{\gamma'}\sigma^K_{\gamma'\gamma\beta}(\alpha)\cdot\exp\left[i\vec{K}_{\gamma'}\cdot(\vec{R}+\vec{A}_u-\vec{v}_{\gamma'})\right].$$

Using this and (21.18), we get:

$$= \sum_{Ru\gamma}\sum_{p_c q_a q_c}K^o(\gamma cp_c,K\gamma'\epsilon e,aq_a)D^{oc}_{q_c p_c}(\alpha)\langle cp_c|\vec{k}_j\beta bp_b\rangle\exp\left[i\vec{K}_{\gamma'}\cdot(\vec{R}+\vec{A}_u-\vec{v}_{\gamma'})\right]$$
$$\cdot\exp\left[i\alpha\vec{K}_j\cdot(\vec{R}+\vec{A}_u-\vec{a})\right]\langle\vec{r}-\vec{R}-\vec{A}_u|\varphi_a aq_a\rangle$$

and further with (21.20):

$$= \sum_{Ru\gamma iq_a q_b q_c}\sum K^o(\gamma cq_c,K\gamma'\epsilon e,aq_a)\cdot D^{\vec{k}b}_{q_b p_b}(\alpha_i^{-1}\alpha\alpha_j)\Delta^{\vec{k}}_{ij}(\alpha)$$
$$\cdot\exp\left[i\vec{K}_{\gamma'}\cdot(\vec{R}+\vec{A}_u-\vec{v}_{\gamma'})\right]\exp\left[i\alpha\vec{K}_j\cdot(\vec{R}+\vec{A}_u-\vec{a})\right]\langle\vec{r}-\vec{R}-\vec{A}_u|\varphi_a aq_a\rangle$$

From $\Delta^{\vec{k}}_{ij}(\alpha)\neq0$ follows again $\alpha\vec{k}_j=\vec{k}_i+\vec{K}'_i$, i.e. for internal points $\alpha\vec{k}_j=\vec{k}_i$. This requirement restricts the method to internal points. We no collect the sums according to (21.10 and 23):

$$= \sum_{iq_b}D^{(^+kb)}_{iq_b,jp_b}(\{\alpha|\vec{a}\})\langle\vec{r}|TB\eta(^+kb)iq_b\rangle$$

and this had to be shown.

We have pointed out, how we principally can determine the s.-a. functions and thereby the SALC coefficients of space groups. A more systematic approach results, if we derive, from the symmetrized plain waves, the standard functions of the space groups as in section 12. This yields complete orthonormal sets of SALC coefficients.

There still remains the problem of the surface of the Brilluoin zone. In individual cases, one may obtain symmetrized plain waves or tight-binding functions by induction from a symmorphic subgroup as shown in [37].

But the systematic approach requires the projective representations of the little cogroups. As has been shown in [71], also the Clebsch-Gordan coefficients or 3jm symbols of the space groups are related to those of the projective representations of the little cogroups. They are treated systematically in the paper of Dirl [70]. Starting from the references [69-71], one can elaborate the Wigner-Racah algebra, then the polyhedral invariants of the space groups, and apply both in the theorems demonstrated in this treatise.

22. Case study: Tetrahedral structures

22.1. Preliminaries and standard functions

As an example, we consider the polyhedral symmetry properties of some tetrahedral structures like the molecules P_4 and P_4O_6. The basis of this consideration are the classic tables of the characters, the 3jm, the 6j and the 9j symbols of the group T_d. Because of the isomorphism of the point groups T_d and O, we can take over the tables given by Griffith [5] for the octahedral group O. The full matrices of the irreducible representations have been tabulated by McWeeny [30], table 4.19/20. The operation of the group elements of T_d on the position vector $\vec{r}$ is listed in the following table.

Table 1. The elements of T_d and their operation on $\vec{r}$.

g	x y z	g	x y z	g	x y z	g	x y z
E	x y z	$C_3^{\bar{x}yz}$	z-x-y	$\bar{S}_4^z$	y-x-z	S_4^x	-x-z y
C_2^x	x-y-z	$C_3^{\bar{x}yz}$	-z x-y	σ_{xy}	-y-x z	$\sigma_{y\bar{z}}$	x z y
C_2^y	-x y-z	$\bar{C}_3^{xyz}$	y z x	$\sigma_{x\bar{y}}$	y x z	S_4^y	z-y-x
C_2^z	-x-y z	$\bar{C}_3^{x\bar{y}z}$	-y z-x	S_4^z	-y x-z	$\sigma_{z\bar{x}}$	z y x
C_3^{xyz}	z x y	$\bar{C}_3^{\bar{x}yz}$	-y-z x	σ_{yz}	x-z-y	σ_{zx}	-z y-x
$C_3^{x\bar{y}z}$	-z-x y	$\bar{C}_3^{\bar{x}\bar{y}z}$	y-z-x	$\bar{S}_4^x$	-x z-y	$\bar{S}_4^y$	-z-y x

Before considering specified structures, we have to prepare a complete set of standard functions of the group T_d following section 12. From the compilation by Bell [38] we take the following set of non-orthogonal s.-a. functions being complete in the sense of the scalar product (12.2):

1) species A_1:
$$\langle \vec{r} | A_1 \rangle = 1 \tag{22.1}$$

2) species A_2:
$$\langle \vec{r} | A_2 \rangle = (x^2-y^2)(y^2-z^2)(z^2-x^2) \tag{22.2}$$

3) species E (components 1 and 2):
$$\langle \vec{r} | 1E1 \rangle = 2z^2-x^2-y^2 \qquad \langle \vec{r} | 2E1 \rangle = 2z^4-x^4-y^4$$
$$\langle \vec{r} | 1E2 \rangle = \sqrt{3}(x^2-y^2) \qquad \langle \vec{r} | 2E2 \rangle = \sqrt{3}(x^4-y^4) \tag{22.3}$$

4) species T_1 (components ξ, η, ζ):
$$\langle \vec{r} | 1T_1\xi \rangle = (y^2-z^2)x, \ \langle \vec{r} | 2T_1\xi \rangle = (y^2-z^2)yz, \ \langle \vec{r} | 3T_1\xi \rangle = (y^2-z^2)x^3$$
$$\langle \vec{r} | 1T_1\eta \rangle = (z^2-x^2)y, \ \langle \vec{r} | 2T_1\eta \rangle = (z^2-x^2)zx, \ \langle \vec{r} | 3T_1\eta \rangle = (z^2-x^2)y^3 \tag{22.4}$$
$$\langle \vec{r} | 1T_1\zeta \rangle = (x^2-y^2)z, \ \langle \vec{r} | 2T_1\zeta \rangle = (x^2-y^2)xy, \ \langle \vec{r} | 3T_1\zeta \rangle = (x^2-y^2)z^3$$

5) species T_2 (components x, y, z):
$$\langle \vec{r} | 1T_2x \rangle = x \qquad \langle \vec{r} | 2T_2x \rangle = yz \qquad \langle \vec{r} | 3T_2x \rangle = x^3$$
$$\langle \vec{r} | 1T_2y \rangle = y \qquad \langle \vec{r} | 2T_2y \rangle = zx \qquad \langle \vec{r} | 3T_2y \rangle = y^3 \tag{22.5}$$
$$\langle \vec{r} | 1T_2z \rangle = z \qquad \langle \vec{r} | 2T_2z \rangle = xy \qquad \langle \vec{r} | 3T_2z \rangle = z^3$$

Because the elements of the Gram matrix (12.9) are scalar functions
of $\vec{r}$, it is advisable to study the scalar functions with respect to T_d
in some detail. There are even scalars of order 2n:

$$Sc(2n) = x^{2n} + y^{2n} + z^{2n} \tag{22.6}$$

and a second type of order 2m+2n:

$$Sc(2m,2n) = y^{2m}z^{2n} + z^{2m}y^{2n} + z^{2m}x^{2n} + x^{2m}z^{2n} + x^{2m}y^{2n} + y^{2m}x^{2n} \tag{22.7}$$

The latter type includes the special case:

$$Sc(2n,2n) = 2(y^{2n}z^{2n} + z^{2n}x^{2n} + x^{2n}y^{2n})$$

A third type of even scalar functions is:

$$Sc(2m,2n,2p) = \sum_{i \neq j \neq k} x_i^{2m} x_j^{2n} x_k^{2p} \tag{22.8}$$

including the special case:

$$Sc(2p,2p,2p) = 6x^{2p}y^{2p}z^{2p}$$

The only essential odd scalar is

$$Sc_o = x \cdot y \cdot z, \tag{22.9}$$

from which all other odd scalars result by multiplication by an even
scalar.

Also the even scalars are not independent, but can be reduced to the
three basic scalars $Sc(2) = r^2$, Sc_o, and $Sc(2,2) = 2(y^2 z^2 + z^2 x^2 + x^2 y^2)$.
We give some examples:

$$Sc(4) = x^4 + y^4 + z^4 = Sc(2)^2 - Sc(2,2)$$
$$Sc(4,2) = 0.5 Sc(2) \cdot Sc(2,2) - 3 Sc_o^2$$
$$Sc(6) = Sc(2)^3 - 1.5 Sc(2) \cdot Sc(2,2) + 3 Sc_o^2$$

We now build up the system of standard function beginning with Gram's
matrix according to (12.9):

$$S_{mn}^a(\vec{r}) = (\text{map}|\text{nap}) = \sum_{g \in G} \langle \text{map}|g\vec{r} \rangle \langle g\vec{r}|\text{nap} \rangle = \text{dima}^{-1} \sum_{g \in G} \sum_p \langle \text{map}|g\vec{r} \rangle \langle g\vec{r}|\text{nap} \rangle$$
$$= (\text{ordG}/\text{dima}) \sum_p \langle \text{map}|\vec{r} \rangle \langle \vec{r}|\text{nap} \rangle \tag{22.10}$$

The evaluation of the last sum in general requires fewer terms. The
first standard function can be taken directly from the sets (22.1-5):

$$\langle \vec{r}|\text{st.1ap} \rangle = \langle \vec{r}|\text{1ap} \rangle \tag{22.11}$$

with $\mu(1a,\vec{r}) = S_{11}^a(\vec{r})$ according to (12.3).
In the case of two- and three-dimensional representations, we slightly
modify Schmidt's orthogonalization process in order to generate purely
polynomial functions:

$$\langle \vec{r}|\text{st.2ap} \rangle = S_{11}^a(\vec{r}) \langle \vec{r}|\text{2ap} \rangle - S_{12}^a(\vec{r}) \langle \vec{r}|\text{1ap} \rangle \tag{22.12}$$

with

$$\mu(2a,\vec{r}) = (S_{11}^a)^2 \cdot S_{22}^a - S_{11}^a \cdot (S_{12}^a)^2 = \mu(1a,\vec{r}) \cdot \det \begin{vmatrix} S_{11}^a & S_{12}^a \\ S_{21}^a & S_{22}^a \end{vmatrix} \tag{22.13}$$

In the case of the three-dimensional representations, the third standard function is given by:

$$\langle \vec{r}|\text{st.3ap}\rangle \tag{22.14}$$
$$= (S_{11}^a S_{22}^a - (S_{12}^a)^2)\langle \vec{r}|3\text{ap}\rangle - (S_{11}^a S_{23}^a - S_{12}^a S_{13}^a)\langle \vec{r}|2\text{ap}\rangle - (S_{13}^a S_{22}^a - S_{23}^a S_{12}^a)\langle \vec{r}|1\text{ap}\rangle$$

with

$$\mu(3a,\vec{r}) = \det\begin{vmatrix} S_{11}^a & S_{12}^a \\ S_{21}^a & S_{22}^a \end{vmatrix} \cdot \det\begin{vmatrix} S_{11}^a & S_{12}^a & S_{13}^a \\ S_{21}^a & S_{22}^a & S_{23}^a \\ S_{31}^a & S_{32}^a & S_{33}^a \end{vmatrix} \tag{22.15}$$

The arrangement is such, that the scalar of the highest order $S_{33}^a(\vec{r})$ does not appear in the third standard function (22.14).

We now compile the standard functions of the separate symmetry species:

1) species A_1:
$$\langle \vec{r}|\text{st.}A_1 1\rangle = 1 \quad , \quad \mu(A_1,\vec{r}) = 24 \tag{22.16}$$

2) species A_2:
$$\langle \vec{r}|\text{st.}A_2 1\rangle = (x^2-y^2)(y^2-z^2)(z^2-x^2), \quad \mu(A_2,\vec{r}) = 24(x^2-y^2)^2(y^2-z^2)^2(z^2-x^2)^2 \tag{22.17}$$

3) species E: From (22.3) results:
$$S_{11}^E = 24(2\text{Sc}(4)-\text{Sc}(2,2)), \quad S_{12}^E = 24(2\text{Sc}(6)-\text{Sc}(4,2)), \quad S_{22}^E = 24(2\text{Sc}(8)-\text{Sc}(4,4)) \tag{22.18}$$

and further:

$$\left.\begin{array}{l} \langle \vec{r}|\text{st.1E1}\rangle = 2z^2-x^2-y^2, \quad \langle \vec{r}|\text{st.1E2}\rangle = \sqrt{3}(x^2-y^2) \\[2mm] \quad\quad \mu(1E,\vec{r}) = 24(2\text{Sc}(4) - \text{Sc}(2,2)) \end{array}\right\} \tag{22.19}$$

$$\left.\begin{array}{l} \langle \vec{r}|\text{st.2E1}\rangle = S_{11}^E(\vec{r})\cdot(2z^4-x^4-y^4) - S_{12}^E(\vec{r})\cdot(2z^2-x^2-y^2) \\[2mm] \langle \vec{r}|\text{st.2E2}\rangle = S_{11}^E(\vec{r})\cdot\sqrt{3}(x^4-y^4) - S_{12}^E(\vec{r})\cdot\sqrt{3}(x^2-y^2) \\[2mm] \quad\quad \mu(2E,\vec{r}) = (S_{11}^E)^2\cdot S_{22}^E - S_{11}^E\cdot(S_{12}^E)^2 \end{array}\right\} \tag{22.20}$$

4) species T_1: From (22.4) results:

$$S_{11}^{T_1}(\vec{r}) = 8(\text{Sc}(4,2)-\text{Sc}(2,2,2)), \quad S_{12}^{T_1}(\vec{r}) = 8(2\text{Sc}(4)-\text{Sc}(2,2))\text{Sc}_0$$
$$S_{22}^{T_1}(\vec{r}) = 8(\text{Sc}(6,2)-\text{Sc}(4,4)) \quad , \quad S_{13}^{T_1}(\vec{r}) = 8(\text{Sc}(4,4)-\text{Sc}(4,2,2)) \tag{22.21}$$
$$S_{33}^{T_1}(\vec{r}) = 8(\text{Sc}(6,4)-\text{Sc}(6,2,2)), \quad S_{23}^{T_1}(\vec{r}) = 8(\text{Sc}(4,2)-\text{Sc}(2,2,2))\text{Sc}_0$$

and further:

$$\left.\begin{array}{l} \langle \vec{r}|\text{st.1}T_1 5\rangle = (y^2-z^2)x \quad \text{and cyclic permutations} \\[2mm] \quad\quad \mu(1T_1,\vec{r}) = 8(\text{Sc}(4,2) - \text{Sc}(2,2,2)) \end{array}\right\} \tag{22.22}$$

$$\left.\begin{array}{l} \langle \vec{r}|\text{st.2}T_1 5\rangle = S_{11}^{T_1}(\vec{r})\cdot(y^2-z^2)yz - S_{12}^{T_1}(\vec{r})\cdot(y^2-z^2)x \quad \text{cyclic} \\[2mm] \quad\quad \mu(2T_1,\vec{r}) = (S_{11}^{T_1})^2\cdot S_{22}^{T_1} - S_{11}^{T_1}\cdot(S_{12}^{T_1})^2 \end{array}\right\} \tag{22.23}$$

$$\left.\begin{array}{l} \langle \vec{r}|\text{st.3}T_1 5\rangle = (S_{11}^{T_1}S_{22}^{T_1}-(S_{12}^{T_1})^2)\cdot(y^2-z^2)x^3 - (S_{11}^{T_1}S_{23}^{T_1}-S_{12}^{T_1}S_{13}^{T_1})\cdot(y^2-z^2)yz \\[2mm] \quad\quad -(S_{13}^{T_1}S_{22}^{T_1}-S_{23}^{T_1}S_{12}^{T_1})\cdot(y^2-z^2)x \quad\quad \text{cyclic} \\[2mm] \quad\quad \mu(3T_1,\vec{r}) = \text{formula (22.15)} \end{array}\right\} \tag{22.24}$$

5) species T_2: From (22.5) results:

$$S_{11}^{T_2}(\vec{r}) = 8r^2 \qquad S_{22}^{T_2}(\vec{r}) = 4Sc(2,2) \qquad S_{33}^{T_2}(\vec{r}) = 8Sc(6) \left.\rule{0pt}{12pt}\right\}$$
$$S_{12}^{T_2}(\vec{r}) = 24xyz \qquad S_{13}^{T_2}(\vec{r}) = 8Sc(4) \qquad S_{23}^{T_2}(\vec{r}) = 8r^2 xyz \quad (22.25)$$

and further:

$$\langle\vec{r}|st.1T_2 x\rangle = x \quad \text{and cyclic permutations} \left.\rule{0pt}{22pt}\right\}(22.26)$$
$$\mu(1T_2,\vec{r}) = 8r^2$$

$$\langle\vec{r}|st.2T_2 x\rangle = 8r^2 yz - 24xyz\cdot x \quad \text{cyclic} \left.\rule{0pt}{22pt}\right\}(22.27)$$
$$\mu(2T_2,\vec{r}) = (S_{11}^{T_2})^2\cdot S_{22}^{T_2} - S_{11}^{T_2}\cdot(S_{12}^{T_2})^2$$

$$\langle\vec{r}|st.3T_2 x\rangle = (S_{11}^{T_2}S_{22}^{T_2}-(S_{12}^{T_2})^2)x^3 - (S_{11}^{T_2}S_{23}^{T_2}-S_{12}^{T_2}S_{13}^{T_2})yz$$
$$- (S_{13}^{T_2}S_{22}^{T_2}-S_{23}^{T_2}S_{12}^{T_2})x \quad \text{cyclic} \left.\rule{0pt}{30pt}\right\}(22.28)$$
$$\mu(3T_2,\vec{r}) = \text{formula (22.15)}$$

22.2. Equivalent sets and their SALC coefficients

After this general preparation, we come to the particular structures having T_d symmetry. There are five different sets of equivalent positions or other equivalent objects. These sets are:

1) The central position invariant to all symmetry operations $\vec{0} = (0,0,0)$
2) Four positions $\vec{A}_r$ on the three-fold rotation axes. These and the following positions are given in table 2.
3) Six positions $\vec{B}_r$ on the two-fold rotation axes.
4) Twelve equivalent positions $\vec{C}_r$ on the reflection planes.
5) 24 general positions $\vec{D}_r$ apart from any element of symmetry.

The position vectors in the latter two cases are not determined uniquely. For instance, $\vec{C}_1 = (b,c,c)$ with $b\neq c$ and $2c^2 + b^2 = 1$ would do. For the purpose of numerical calculation, we have arbitrarily chosen the vectors given in table 2.

Table 2. The position vectors of the equivalent sets.

	x	y	z		x	y	z
A_1	$1/\sqrt{3}$	$1/\sqrt{3}$	$1/\sqrt{3}$	C_2	$1/2$	$1/\sqrt{2}$	$1/2$
A_2	$1/\sqrt{3}$	$-1/\sqrt{3}$	$-1/\sqrt{3}$	C_3	$1/2$	$1/2$	$1/\sqrt{2}$
A_3	$-1/\sqrt{3}$	$-1/\sqrt{3}$	$1/\sqrt{3}$	C_4	$1/\sqrt{2}$	$-1/2$	$-1/2$
A_4	$-1/\sqrt{3}$	$1/\sqrt{3}$	$-1/\sqrt{3}$	C_5	$-1/2$	$1/\sqrt{2}$	$-1/2$
B_1	0	0	1	C_6	$-1/2$	$-1/2$	$1/\sqrt{2}$
B_2	0	1	0	C_7	$-1/\sqrt{2}$	$1/2$	$-1/2$
B_3	1	0	0	C_8	$-1/2$	$-1/\sqrt{2}$	$1/2$
B_4	0	0	-1	C_9	$1/2$	$-1/2$	$-1/\sqrt{2}$
B_5	0	-1	0	C_{10}	$-1/\sqrt{2}$	$-1/2$	$1/2$
B_6	-1	0	0	C_{11}	$1/2$	$-1/\sqrt{2}$	$-1/2$
C_1	$1/\sqrt{2}$	$1/2$	$1/2$	C_{12}	$-1/2$	$1/2$	$-1/\sqrt{2}$

Table 2. (continued)

	x	y	z		x	y	z
D_1	1/3	$\sqrt{3}/3$	$\sqrt{5}/3$	D_{13}	$\sqrt{3}/3$	-1/3	$-\sqrt{5}/3$
D_2	1/3	$-\sqrt{3}/3$	$-\sqrt{5}/3$	D_{14}	$-\sqrt{3}/3$	-1/3	$\sqrt{5}/3$
D_3	-1/3	$\sqrt{3}/3$	$-\sqrt{5}/3$	D_{15}	$\sqrt{3}/3$	1/3	$\sqrt{5}/3$
D_4	-1/3	$-\sqrt{3}/3$	$\sqrt{5}/3$	D_{16}	$-\sqrt{3}/3$	1/3	$-\sqrt{5}/3$
D_5	$\sqrt{5}/3$	1/3	$\sqrt{3}/3$	D_{17}	1/3	$-\sqrt{5}/3$	$-\sqrt{3}/3$
D_6	$-\sqrt{5}/3$	-1/3	$\sqrt{3}/3$	D_{18}	-1/3	$\sqrt{5}/3$	$-\sqrt{3}/3$
D_7	$\sqrt{5}/3$	-1/3	$-\sqrt{3}/3$	D_{19}	-1/3	$-\sqrt{5}/3$	$\sqrt{3}/3$
D_8	$-\sqrt{5}/3$	1/3	$-\sqrt{3}/3$	D_{20}	1/3	$\sqrt{5}/3$	$\sqrt{3}/3$
D_9	$\sqrt{3}/3$	$\sqrt{5}/3$	1/3	D_{21}	$\sqrt{5}/3$	$-\sqrt{3}/3$	-1/3
D_{10}	$-\sqrt{3}/3$	$\sqrt{5}/3$	-1/3	D_{22}	$\sqrt{5}/3$	$\sqrt{3}/3$	1/3
D_{11}	$-\sqrt{3}/3$	$-\sqrt{5}/3$	1/3	D_{23}	$-\sqrt{5}/3$	$\sqrt{3}/3$	-1/3
D_{12}	$\sqrt{3}/3$	$-\sqrt{5}/3$	-1/3	D_{24}	$-\sqrt{5}/3$	$-\sqrt{3}/3$	1/3

The enumeration of the vectors of set D corresponds to that of the group elements in table 1.

These five sets of equivalent objects induce the reducible representations $\sigma^0 = A_1$, σ^A, σ^B, σ^C, and σ^D. Using the operations listed in table 1, we can calculate the induced matrices of σ^A etc. according to (3.3). The calculation of the characters is even simpler, since they are equal to the number of positions being invariant to the operation. The characters of the irreducible and the induced representations are listed in table 3.

Table 3. Characters of representations of T_d

	E	$8C_3$	$3C_2$	$6S_4$	$6\sigma_d$
A_1	1	1	1	1	1
A_2	1	1	1	-1	-1
E	2	-1	2	0	0
T_1	3	0	-1	1	-1
T_2	3	0	-1	-1	1
σ^0	1	1	1	1	1
σ^A	4	1	0	0	2
σ^B	6	0	2	0	2
σ^C	12	0	0	0	2
σ^D	24	0	0	0	0

From the character formula (2.10) then follow the branching rules for the induced representations:

$$\sigma^0 = A_1, \quad \sigma^A = A_1 + T_2, \quad \sigma^B = A_1 + E + T_2, \quad \sigma^C = A_1 + E + T_1 + 2T_2$$
$$\sigma^D = A_1 + A_2 + 2E + 3T_1 + 3T_2$$

(22.29)

The decomposition of the representations according to these rules is achieved by the SALC coefficients. We calculate them by formula (12.16), i.e. by inserting the position vectors of table 2 into the standard functions (22.16-28). The results are listed in the tables 4 to 7.

Table 4.
SALC coefficients $(\vec{A}_1|Aap)$

a p	A_1	A_2	A_3	A_4
$A_1 1$	$\frac{1}{2}$	$\frac{1}{2}$	$\frac{1}{2}$	$\frac{1}{2}$
$T_2 x$	$\frac{1}{2}$	$\frac{1}{2}$	$\frac{-1}{2}$	$\frac{-1}{2}$
$T_2 y$	$\frac{1}{2}$	$\frac{-1}{2}$	$\frac{-1}{2}$	$\frac{1}{2}$
$T_2 z$	$\frac{1}{2}$	$\frac{-1}{2}$	$\frac{1}{2}$	$\frac{-1}{2}$

Table 5.
SALC coefficients $(\vec{B}_1|Bbp)$

b p	B_1	B_2	B_3	B_4	B_5	B_6
$A_1 1$	$\frac{\sqrt{6}}{6}$	$\frac{\sqrt{6}}{6}$	$\frac{\sqrt{6}}{6}$	$\frac{\sqrt{6}}{6}$	$\frac{\sqrt{6}}{6}$	$\frac{\sqrt{6}}{6}$
E 1	$\frac{\sqrt{3}}{3}$	$-\frac{\sqrt{3}}{6}$	$-\frac{\sqrt{3}}{6}$	$\frac{\sqrt{3}}{3}$	$-\frac{\sqrt{3}}{6}$	$-\frac{\sqrt{3}}{6}$
E 2	0	$\frac{-1}{2}$	$\frac{1}{2}$	0	$\frac{-1}{2}$	$\frac{1}{2}$
$T_2 x$	0	0	$\frac{\sqrt{2}}{2}$	0	0	$-\frac{\sqrt{2}}{2}$
$T_2 y$	0	$\frac{\sqrt{2}}{2}$	0	0	$-\frac{\sqrt{2}}{2}$	0
$T_2 z$	$\frac{\sqrt{2}}{2}$	0	0	$-\frac{\sqrt{2}}{2}$	0	0

Table 6. SALC coefficients $(\vec{C}_1|C\gamma cp)$

γc p	C_1	C_2	C_3	C_4	C_5	C_6	C_7	C_8	C_9	C_{10}	C_{11}	C_{12}
$A_1 1$	$\frac{\sqrt{3}}{6}$	$\frac{\sqrt{3}}{6}$	$\frac{\sqrt{3}}{6}$	$\frac{\sqrt{3}}{6}$	$\frac{\sqrt{3}}{6}$	$\frac{\sqrt{3}}{6}$	$\frac{\sqrt{3}}{6}$	$\frac{\sqrt{3}}{6}$	$\frac{\sqrt{3}}{6}$	$\frac{\sqrt{3}}{6}$	$\frac{\sqrt{3}}{6}$	$\frac{\sqrt{3}}{6}$
E 1	$-\frac{\sqrt{6}}{12}$	$-\frac{\sqrt{6}}{12}$	$\frac{\sqrt{6}}{6}$	$-\frac{\sqrt{6}}{12}$	$-\frac{\sqrt{6}}{12}$	$\frac{\sqrt{6}}{6}$	$-\frac{\sqrt{6}}{12}$	$-\frac{\sqrt{6}}{12}$	$\frac{\sqrt{6}}{6}$	$-\frac{\sqrt{6}}{12}$	$-\frac{\sqrt{6}}{12}$	$\frac{\sqrt{6}}{6}$
E 2	$\frac{\sqrt{2}}{4}$	$-\frac{\sqrt{2}}{4}$	0	$\frac{\sqrt{2}}{4}$	$-\frac{\sqrt{2}}{4}$	0	$\frac{\sqrt{2}}{4}$	$-\frac{\sqrt{2}}{4}$	0	$\frac{\sqrt{2}}{4}$	$-\frac{\sqrt{2}}{4}$	0
$T_1\xi$	0	$\frac{\sqrt{2}}{4}$	$-\frac{\sqrt{2}}{4}$	0	$-\frac{\sqrt{2}}{4}$	$\frac{\sqrt{2}}{4}$	0	$-\frac{\sqrt{2}}{4}$	$-\frac{\sqrt{2}}{4}$	0	$\frac{\sqrt{2}}{4}$	$\frac{\sqrt{2}}{4}$
$T_1\eta$	$-\frac{\sqrt{2}}{4}$	0	$\frac{\sqrt{2}}{4}$	$\frac{\sqrt{2}}{4}$	0	$-\frac{\sqrt{2}}{4}$	$-\frac{\sqrt{2}}{4}$	0	$-\frac{\sqrt{2}}{4}$	$\frac{\sqrt{2}}{4}$	0	$\frac{\sqrt{2}}{4}$
$T_1\zeta$	$\frac{\sqrt{2}}{4}$	$-\frac{\sqrt{2}}{4}$	0	$-\frac{\sqrt{2}}{4}$	$\frac{\sqrt{2}}{4}$	0	$-\frac{\sqrt{2}}{4}$	$-\frac{\sqrt{2}}{4}$	0	$\frac{\sqrt{2}}{4}$	$\frac{\sqrt{2}}{4}$	0
$1T_2 x$	$\frac{\sqrt{2}}{4}$	$\frac{1}{4}$	$\frac{1}{4}$	$\frac{\sqrt{2}}{4}$	$-\frac{1}{4}$	$-\frac{1}{4}$	$-\frac{\sqrt{2}}{4}$	$-\frac{1}{4}$	$\frac{1}{4}$	$-\frac{\sqrt{2}}{4}$	$\frac{1}{4}$	$-\frac{1}{4}$
$1T_2 y$	$\frac{1}{4}$	$\frac{\sqrt{2}}{4}$	$\frac{1}{4}$	$-\frac{1}{4}$	$\frac{\sqrt{2}}{4}$	$-\frac{1}{4}$	$\frac{1}{4}$	$-\frac{\sqrt{2}}{4}$	$-\frac{1}{4}$	$-\frac{1}{4}$	$-\frac{\sqrt{2}}{4}$	$\frac{1}{4}$
$1T_2 z$	$\frac{1}{4}$	$\frac{1}{4}$	$\frac{\sqrt{2}}{4}$	$-\frac{1}{4}$	$-\frac{1}{4}$	$\frac{\sqrt{2}}{4}$	$-\frac{1}{4}$	$\frac{1}{4}$	$-\frac{\sqrt{2}}{4}$	$\frac{1}{4}$	$-\frac{1}{4}$	$-\frac{\sqrt{2}}{4}$
$2T_2 x$	$-\frac{\sqrt{2}}{4}$	$\frac{1}{4}$	$\frac{1}{4}$	$-\frac{\sqrt{2}}{4}$	$-\frac{1}{4}$	$-\frac{1}{4}$	$\frac{\sqrt{2}}{4}$	$-\frac{1}{4}$	$\frac{1}{4}$	$\frac{\sqrt{2}}{4}$	$\frac{1}{4}$	$-\frac{1}{4}$
$2T_2 y$	$\frac{1}{4}$	$-\frac{\sqrt{2}}{4}$	$\frac{1}{4}$	$-\frac{1}{4}$	$-\frac{\sqrt{2}}{4}$	$-\frac{1}{4}$	$\frac{1}{4}$	$\frac{\sqrt{2}}{4}$	$-\frac{1}{4}$	$-\frac{1}{4}$	$\frac{\sqrt{2}}{4}$	$\frac{1}{4}$
$2T_2 z$	$\frac{1}{4}$	$\frac{1}{4}$	$-\frac{\sqrt{2}}{4}$	$-\frac{1}{4}$	$-\frac{1}{4}$	$-\frac{\sqrt{2}}{4}$	$-\frac{1}{4}$	$\frac{1}{4}$	$\frac{\sqrt{2}}{4}$	$\frac{1}{4}$	$-\frac{1}{4}$	$\frac{\sqrt{2}}{4}$

Table 7. SALC coefficients $(\vec{D}_i | D\delta dp)$

δd p	D_1	D_2	D_3	D_4	D_5	D_6	D_7	D_8	D_9	D_{10}	D_{11}	D_{12}
$A_1 1$	$\frac{\sqrt6}{12}$	$\frac{\sqrt6}{12}$	$\frac{\sqrt6}{12}$	$\frac{\sqrt6}{12}$	$\frac{\sqrt6}{12}$	$\frac{\sqrt6}{12}$	$\frac{\sqrt6}{12}$	$\frac{\sqrt6}{12}$	$\frac{\sqrt6}{12}$	$\frac{\sqrt6}{12}$	$\frac{\sqrt6}{12}$	$\frac{\sqrt6}{12}$
$A_2 1$	$\frac{\sqrt6}{12}$	$\frac{\sqrt6}{12}$	$\frac{\sqrt6}{12}$	$\frac{\sqrt6}{12}$	$\frac{\sqrt6}{12}$	$\frac{\sqrt6}{12}$	$\frac{\sqrt6}{12}$	$\frac{\sqrt6}{12}$	$\frac{\sqrt6}{12}$	$\frac{\sqrt6}{12}$	$\frac{\sqrt6}{12}$	$\frac{\sqrt6}{12}$
$1E\ 1$	$\frac14$	$\frac14$	$\frac14$	$\frac14$	0	0	0	0	$\frac{-1}{4}$	$\frac{-1}{4}$	$\frac{-1}{4}$	$\frac{-1}{4}$
$1E\ 2$	$\frac{-\sqrt3}{12}$	$\frac{-\sqrt3}{12}$	$\frac{-\sqrt3}{12}$	$\frac{-\sqrt3}{12}$	$\frac{\sqrt3}{6}$	$\frac{\sqrt3}{6}$	$\frac{\sqrt3}{6}$	$\frac{\sqrt3}{6}$	$\frac{-\sqrt3}{12}$	$\frac{-\sqrt3}{12}$	$\frac{-\sqrt3}{12}$	$\frac{-\sqrt3}{12}$
$2E\ 1$	$\frac{\sqrt3}{12}$	$\frac{\sqrt3}{12}$	$\frac{\sqrt3}{12}$	$\frac{\sqrt3}{12}$	$\frac{-\sqrt3}{6}$	$\frac{-\sqrt3}{6}$	$\frac{-\sqrt3}{6}$	$\frac{-\sqrt3}{6}$	$\frac{\sqrt3}{12}$	$\frac{\sqrt3}{12}$	$\frac{\sqrt3}{12}$	$\frac{\sqrt3}{12}$
$2E\ 2$	$\frac14$	$\frac14$	$\frac14$	$\frac14$	0	0	0	0	$\frac{-1}{4}$	$\frac{-1}{4}$	$\frac{-1}{4}$	$\frac{-1}{4}$
$1T_1\xi$	$\frac{-1}{12}$	$\frac{-1}{12}$	$\frac{1}{12}$	$\frac{1}{12}$	$\frac{-\sqrt5}{12}$	$\frac{\sqrt5}{12}$	$\frac{-\sqrt5}{12}$	$\frac{\sqrt5}{12}$	$\frac{\sqrt3}{6}$	$\frac{-\sqrt3}{6}$	$\frac{-\sqrt3}{6}$	$\frac{\sqrt3}{6}$
$1T_1\eta$	$\frac{\sqrt3}{6}$	$\frac{-\sqrt3}{6}$	$\frac{\sqrt3}{6}$	$\frac{-\sqrt3}{6}$	$\frac{-1}{12}$	$\frac{1}{12}$	$\frac{1}{12}$	$\frac{-1}{12}$	$\frac{-\sqrt5}{12}$	$\frac{-\sqrt5}{12}$	$\frac{\sqrt5}{12}$	$\frac{\sqrt5}{12}$
$1T_1\zeta$	$\frac{-\sqrt5}{12}$	$\frac{\sqrt5}{12}$	$\frac{\sqrt5}{12}$	$\frac{-\sqrt5}{12}$	$\frac{\sqrt3}{6}$	$\frac{\sqrt3}{6}$	$\frac{-\sqrt3}{6}$	$\frac{-\sqrt3}{6}$	$\frac{-1}{12}$	$\frac{1}{12}$	$\frac{-1}{12}$	$\frac{1}{12}$
$2T_1\xi$	$\frac{-\sqrt{15}}{12}$	$\frac{-\sqrt{15}}{12}$	$\frac{\sqrt{15}}{12}$	$\frac{\sqrt{15}}{12}$	$\frac{\sqrt3}{12}$	$\frac{-\sqrt3}{12}$	$\frac{\sqrt3}{12}$	$\frac{-\sqrt3}{12}$	0	0	0	0
$2T_1\eta$	0	0	0	0	$\frac{-\sqrt{15}}{12}$	$\frac{\sqrt{15}}{12}$	$\frac{\sqrt{15}}{12}$	$\frac{-\sqrt{15}}{12}$	$\frac{\sqrt3}{12}$	$\frac{\sqrt3}{12}$	$\frac{-\sqrt3}{12}$	$\frac{-\sqrt3}{12}$
$2T_1\zeta$	$\frac{\sqrt3}{12}$	$\frac{-\sqrt3}{12}$	$\frac{-\sqrt3}{12}$	$\frac{\sqrt3}{12}$	0	0	0	0	$\frac{-\sqrt{15}}{12}$	$\frac{\sqrt{15}}{12}$	$\frac{-\sqrt{15}}{12}$	$\frac{\sqrt{15}}{12}$
$3T_1\xi$	$\frac{-\sqrt2}{12}$	$\frac{-\sqrt2}{12}$	$\frac{\sqrt2}{12}$	$\frac{\sqrt2}{12}$	$\frac{-\sqrt{10}}{12}$	$\frac{\sqrt{10}}{12}$	$\frac{-\sqrt{10}}{12}$	$\frac{\sqrt{10}}{12}$	$\frac{-\sqrt6}{12}$	$\frac{\sqrt6}{12}$	$\frac{\sqrt6}{12}$	$\frac{-\sqrt6}{12}$
$3T_1\eta$	$\frac{-\sqrt6}{12}$	$\frac{\sqrt6}{12}$	$\frac{-\sqrt6}{12}$	$\frac{\sqrt6}{12}$	$\frac{-\sqrt2}{12}$	$\frac{\sqrt2}{12}$	$\frac{\sqrt2}{12}$	$\frac{-\sqrt2}{12}$	$\frac{-\sqrt{10}}{12}$	$\frac{-\sqrt{10}}{12}$	$\frac{\sqrt{10}}{12}$	$\frac{\sqrt{10}}{12}$
$3T_1\zeta$	$\frac{-\sqrt{10}}{12}$	$\frac{\sqrt{10}}{12}$	$\frac{\sqrt{10}}{12}$	$\frac{-\sqrt{10}}{12}$	$\frac{-\sqrt6}{12}$	$\frac{-\sqrt6}{12}$	$\frac{\sqrt6}{12}$	$\frac{\sqrt6}{12}$	$\frac{-\sqrt2}{12}$	$\frac{\sqrt2}{12}$	$\frac{-\sqrt2}{12}$	$\frac{\sqrt2}{12}$
$1T_2 x$	$\frac{\sqrt2}{12}$	$\frac{\sqrt2}{12}$	$\frac{-\sqrt2}{12}$	$\frac{-\sqrt2}{12}$	$\frac{\sqrt{10}}{12}$	$\frac{-\sqrt{10}}{12}$	$\frac{\sqrt{10}}{12}$	$\frac{-\sqrt{10}}{12}$	$\frac{\sqrt6}{12}$	$\frac{-\sqrt6}{12}$	$\frac{-\sqrt6}{12}$	$\frac{\sqrt6}{12}$
$1T_2 y$	$\frac{\sqrt6}{12}$	$\frac{-\sqrt6}{12}$	$\frac{\sqrt6}{12}$	$\frac{-\sqrt6}{12}$	$\frac{\sqrt2}{12}$	$\frac{-\sqrt2}{12}$	$\frac{-\sqrt2}{12}$	$\frac{\sqrt2}{12}$	$\frac{\sqrt{10}}{12}$	$\frac{\sqrt{10}}{12}$	$\frac{-\sqrt{10}}{12}$	$\frac{-\sqrt{10}}{12}$
$1T_2 z$	$\frac{\sqrt{10}}{12}$	$\frac{-\sqrt{10}}{12}$	$\frac{-\sqrt{10}}{12}$	$\frac{\sqrt{10}}{12}$	$\frac{\sqrt6}{12}$	$\frac{\sqrt6}{12}$	$\frac{-\sqrt6}{12}$	$\frac{-\sqrt6}{12}$	$\frac{\sqrt2}{12}$	$\frac{-\sqrt2}{12}$	$\frac{\sqrt2}{12}$	$\frac{-\sqrt2}{12}$
$2T_2 x$	$\frac{\sqrt{15}}{12}$	$\frac{\sqrt{15}}{12}$	$\frac{-\sqrt{15}}{12}$	$\frac{-\sqrt{15}}{12}$	$\frac{-\sqrt3}{12}$	$\frac{\sqrt3}{12}$	$\frac{-\sqrt3}{12}$	$\frac{\sqrt3}{12}$	0	0	0	0
$2T_2 y$	0	0	0	0	$\frac{\sqrt{15}}{12}$	$\frac{-\sqrt{15}}{12}$	$\frac{-\sqrt{15}}{12}$	$\frac{\sqrt{15}}{12}$	$\frac{-\sqrt3}{12}$	$\frac{-\sqrt3}{12}$	$\frac{\sqrt3}{12}$	$\frac{\sqrt3}{12}$
$2T_2 z$	$\frac{-\sqrt3}{12}$	$\frac{\sqrt3}{12}$	$\frac{\sqrt3}{12}$	$\frac{-\sqrt3}{12}$	0	0	0	0	$\frac{\sqrt{15}}{12}$	$\frac{-\sqrt{15}}{12}$	$\frac{\sqrt{15}}{12}$	$\frac{-\sqrt{15}}{12}$
$3T_2 x$	$\frac{1}{12}$	$\frac{1}{12}$	$\frac{-1}{12}$	$\frac{-1}{12}$	$\frac{\sqrt5}{12}$	$\frac{-\sqrt5}{12}$	$\frac{\sqrt5}{12}$	$\frac{-\sqrt5}{12}$	$\frac{-\sqrt3}{6}$	$\frac{\sqrt3}{6}$	$\frac{\sqrt3}{6}$	$\frac{-\sqrt3}{6}$
$3T_2 y$	$\frac{-\sqrt3}{6}$	$\frac{\sqrt3}{6}$	$\frac{-\sqrt3}{6}$	$\frac{\sqrt3}{6}$	$\frac{1}{12}$	$\frac{-1}{12}$	$\frac{-1}{12}$	$\frac{1}{12}$	$\frac{\sqrt5}{12}$	$\frac{\sqrt5}{12}$	$\frac{-\sqrt5}{12}$	$\frac{-\sqrt5}{12}$
$3T_2 z$	$\frac{\sqrt5}{12}$	$\frac{-\sqrt5}{12}$	$\frac{-\sqrt5}{12}$	$\frac{\sqrt5}{12}$	$\frac{-\sqrt3}{6}$	$\frac{-\sqrt3}{6}$	$\frac{\sqrt3}{6}$	$\frac{\sqrt3}{6}$	$\frac{1}{12}$	$\frac{-1}{12}$	$\frac{1}{12}$	$\frac{-1}{12}$

Table 7. (continued)

$\delta d\ p$	D_{13}	D_{14}	D_{15}	D_{16}	D_{17}	D_{18}	D_{19}	D_{20}	D_{21}	D_{22}	D_{23}	D_{24}
$A_1 1$	$\frac{\sqrt{6}}{12}$	$\frac{\sqrt{6}}{12}$	$\frac{\sqrt{6}}{12}$	$\frac{\sqrt{6}}{12}$	$\frac{\sqrt{6}}{12}$	$\frac{\sqrt{6}}{12}$	$\frac{\sqrt{6}}{12}$	$\frac{\sqrt{6}}{12}$	$\frac{\sqrt{6}}{12}$	$\frac{\sqrt{6}}{12}$	$\frac{\sqrt{6}}{12}$	$\frac{\sqrt{6}}{12}$
$A_2 1$	$-\frac{\sqrt{6}}{12}$	$-\frac{\sqrt{6}}{12}$	$-\frac{\sqrt{6}}{12}$	$-\frac{\sqrt{6}}{12}$	$-\frac{\sqrt{6}}{12}$	$-\frac{\sqrt{6}}{12}$	$-\frac{\sqrt{6}}{12}$	$-\frac{\sqrt{6}}{12}$	$-\frac{\sqrt{6}}{12}$	$-\frac{\sqrt{6}}{12}$	$-\frac{\sqrt{6}}{12}$	$-\frac{\sqrt{6}}{12}$
$1E\,1$	$\frac{1}{4}$	$\frac{1}{4}$	$\frac{1}{4}$	$\frac{1}{4}$	0	0	0	0	$-\frac{1}{4}$	$-\frac{1}{4}$	$-\frac{1}{4}$	$-\frac{1}{4}$
$1E\,2$	$\frac{\sqrt{3}}{12}$	$\frac{\sqrt{3}}{12}$	$\frac{\sqrt{3}}{12}$	$\frac{\sqrt{3}}{12}$	$-\frac{\sqrt{3}}{6}$	$-\frac{\sqrt{3}}{6}$	$-\frac{\sqrt{3}}{6}$	$-\frac{\sqrt{3}}{6}$	$\frac{\sqrt{3}}{12}$	$\frac{\sqrt{3}}{12}$	$\frac{\sqrt{3}}{12}$	$\frac{\sqrt{3}}{12}$
$2E\,1$	$\frac{\sqrt{3}}{12}$	$\frac{\sqrt{3}}{12}$	$\frac{\sqrt{3}}{12}$	$\frac{\sqrt{3}}{12}$	$-\frac{\sqrt{3}}{6}$	$-\frac{\sqrt{3}}{6}$	$-\frac{\sqrt{3}}{6}$	$-\frac{\sqrt{3}}{6}$	$\frac{\sqrt{3}}{12}$	$\frac{\sqrt{3}}{12}$	$\frac{\sqrt{3}}{12}$	$\frac{\sqrt{3}}{12}$
$2E\,2$	$-\frac{1}{4}$	$-\frac{1}{4}$	$-\frac{1}{4}$	$-\frac{1}{4}$	0	0	0	0	$\frac{1}{4}$	$\frac{1}{4}$	$\frac{1}{4}$	$\frac{1}{4}$
$1T_1\xi$	$-\frac{\sqrt{3}}{6}$	$\frac{\sqrt{3}}{6}$	$-\frac{\sqrt{3}}{6}$	$\frac{\sqrt{3}}{6}$	$\frac{1}{12}$	$-\frac{1}{12}$	$-\frac{1}{12}$	$\frac{1}{12}$	$\frac{\sqrt{5}}{12}$	$\frac{\sqrt{5}}{12}$	$-\frac{\sqrt{5}}{12}$	$-\frac{\sqrt{5}}{12}$
$1T_1\eta$	$-\frac{1}{12}$	$-\frac{1}{12}$	$\frac{1}{12}$	$\frac{1}{12}$	$-\frac{\sqrt{5}}{12}$	$\frac{\sqrt{5}}{12}$	$-\frac{\sqrt{5}}{12}$	$\frac{\sqrt{5}}{12}$	$\frac{\sqrt{3}}{6}$	$-\frac{\sqrt{3}}{6}$	$-\frac{\sqrt{3}}{6}$	$\frac{\sqrt{3}}{6}$
$1T_1\zeta$	$-\frac{\sqrt{5}}{12}$	$\frac{\sqrt{5}}{12}$	$\frac{\sqrt{5}}{12}$	$-\frac{\sqrt{5}}{12}$	$\frac{\sqrt{3}}{6}$	$\frac{\sqrt{3}}{6}$	$-\frac{\sqrt{3}}{6}$	$-\frac{\sqrt{3}}{6}$	$-\frac{1}{12}$	$\frac{1}{12}$	$-\frac{1}{12}$	$\frac{1}{12}$
$2T_1\xi$	0	0	0	0	$\frac{\sqrt{15}}{12}$	$-\frac{\sqrt{15}}{12}$	$-\frac{\sqrt{15}}{12}$	$\frac{\sqrt{15}}{12}$	$-\frac{\sqrt{3}}{12}$	$-\frac{\sqrt{3}}{12}$	$\frac{\sqrt{3}}{12}$	$\frac{\sqrt{3}}{12}$
$2T_1\eta$	$-\frac{\sqrt{15}}{12}$	$-\frac{\sqrt{15}}{12}$	$\frac{\sqrt{15}}{12}$	$\frac{\sqrt{15}}{12}$	$\frac{\sqrt{3}}{12}$	$-\frac{\sqrt{3}}{12}$	$\frac{\sqrt{3}}{12}$	$-\frac{\sqrt{3}}{12}$	0	0	0	0
$2T_1\zeta$	$\frac{\sqrt{3}}{12}$	$-\frac{\sqrt{3}}{12}$	$-\frac{\sqrt{3}}{12}$	$\frac{\sqrt{3}}{12}$	0	0	0	0	$-\frac{\sqrt{15}}{12}$	$\frac{\sqrt{15}}{12}$	$-\frac{\sqrt{15}}{12}$	$\frac{\sqrt{15}}{12}$
$3T_1\xi$	$\frac{\sqrt{6}}{12}$	$-\frac{\sqrt{6}}{12}$	$\frac{\sqrt{6}}{12}$	$-\frac{\sqrt{6}}{12}$	$-\frac{\sqrt{2}}{12}$	$\frac{\sqrt{2}}{12}$	$\frac{\sqrt{2}}{12}$	$-\frac{\sqrt{2}}{12}$	$\frac{\sqrt{10}}{12}$	$\frac{\sqrt{10}}{12}$	$-\frac{\sqrt{10}}{12}$	$-\frac{\sqrt{10}}{12}$
$3T_1\eta$	$-\frac{\sqrt{2}}{12}$	$-\frac{\sqrt{2}}{12}$	$\frac{\sqrt{2}}{12}$	$\frac{\sqrt{2}}{12}$	$-\frac{\sqrt{10}}{12}$	$\frac{\sqrt{10}}{12}$	$-\frac{\sqrt{10}}{12}$	$\frac{\sqrt{10}}{12}$	$-\frac{\sqrt{6}}{12}$	$\frac{\sqrt{6}}{12}$	$\frac{\sqrt{6}}{12}$	$-\frac{\sqrt{6}}{12}$
$3T_1\zeta$	$-\frac{\sqrt{10}}{12}$	$\frac{\sqrt{10}}{12}$	$\frac{\sqrt{10}}{12}$	$-\frac{\sqrt{10}}{12}$	$-\frac{\sqrt{6}}{12}$	$-\frac{\sqrt{6}}{12}$	$\frac{\sqrt{6}}{12}$	$\frac{\sqrt{6}}{12}$	$-\frac{\sqrt{2}}{12}$	$\frac{\sqrt{2}}{12}$	$-\frac{\sqrt{2}}{12}$	$\frac{\sqrt{2}}{12}$
$1T_2 x$	$\frac{\sqrt{6}}{12}$	$-\frac{\sqrt{6}}{12}$	$\frac{\sqrt{6}}{12}$	$-\frac{\sqrt{6}}{12}$	$\frac{\sqrt{2}}{12}$	$-\frac{\sqrt{2}}{12}$	$-\frac{\sqrt{2}}{12}$	$\frac{\sqrt{2}}{12}$	$\frac{\sqrt{10}}{12}$	$\frac{\sqrt{10}}{12}$	$-\frac{\sqrt{10}}{12}$	$-\frac{\sqrt{10}}{12}$
$1T_2 y$	$-\frac{\sqrt{2}}{12}$	$-\frac{\sqrt{2}}{12}$	$\frac{\sqrt{2}}{12}$	$\frac{\sqrt{2}}{12}$	$-\frac{\sqrt{10}}{12}$	$\frac{\sqrt{10}}{12}$	$-\frac{\sqrt{10}}{12}$	$\frac{\sqrt{10}}{12}$	$-\frac{\sqrt{6}}{12}$	$\frac{\sqrt{6}}{12}$	$\frac{\sqrt{6}}{12}$	$-\frac{\sqrt{6}}{12}$
$1T_2 z$	$-\frac{\sqrt{10}}{12}$	$\frac{\sqrt{10}}{12}$	$\frac{\sqrt{10}}{12}$	$-\frac{\sqrt{10}}{12}$	$-\frac{\sqrt{6}}{12}$	$-\frac{\sqrt{6}}{12}$	$\frac{\sqrt{6}}{12}$	$\frac{\sqrt{6}}{12}$	$-\frac{\sqrt{2}}{12}$	$\frac{\sqrt{2}}{12}$	$-\frac{\sqrt{2}}{12}$	$\frac{\sqrt{2}}{12}$
$2T_2 x$	0	0	0	0	$\frac{\sqrt{15}}{12}$	$-\frac{\sqrt{15}}{12}$	$-\frac{\sqrt{15}}{12}$	$\frac{\sqrt{15}}{12}$	$-\frac{\sqrt{3}}{12}$	$-\frac{\sqrt{3}}{12}$	$\frac{\sqrt{3}}{12}$	$\frac{\sqrt{3}}{12}$
$2T_2 y$	$-\frac{\sqrt{15}}{12}$	$-\frac{\sqrt{15}}{12}$	$\frac{\sqrt{15}}{12}$	$\frac{\sqrt{15}}{12}$	$\frac{\sqrt{3}}{12}$	$-\frac{\sqrt{3}}{12}$	$\frac{\sqrt{3}}{12}$	$-\frac{\sqrt{3}}{12}$	0	0	0	0
$2T_2 z$	$\frac{\sqrt{3}}{12}$	$-\frac{\sqrt{3}}{12}$	$-\frac{\sqrt{3}}{12}$	$\frac{\sqrt{3}}{12}$	0	0	0	0	$-\frac{\sqrt{15}}{12}$	$\frac{\sqrt{15}}{12}$	$-\frac{\sqrt{15}}{12}$	$\frac{\sqrt{15}}{12}$
$3T_2 x$	$-\frac{\sqrt{3}}{6}$	$\frac{\sqrt{3}}{6}$	$\frac{\sqrt{3}}{6}$	$-\frac{\sqrt{3}}{6}$	$\frac{1}{12}$	$-\frac{1}{12}$	$-\frac{1}{12}$	$\frac{1}{12}$	$\frac{\sqrt{5}}{12}$	$\frac{\sqrt{5}}{12}$	$-\frac{\sqrt{5}}{12}$	$-\frac{\sqrt{5}}{12}$
$3T_2 y$	$-\frac{1}{12}$	$-\frac{1}{12}$	$\frac{1}{12}$	$\frac{1}{12}$	$-\frac{\sqrt{5}}{12}$	$\frac{\sqrt{5}}{12}$	$-\frac{\sqrt{5}}{12}$	$\frac{\sqrt{5}}{12}$	$\frac{\sqrt{3}}{6}$	$-\frac{\sqrt{3}}{6}$	$-\frac{\sqrt{3}}{6}$	$\frac{\sqrt{3}}{6}$
$3T_2 z$	$-\frac{\sqrt{5}}{12}$	$\frac{\sqrt{5}}{12}$	$\frac{\sqrt{5}}{12}$	$-\frac{\sqrt{5}}{12}$	$\frac{\sqrt{3}}{6}$	$\frac{\sqrt{3}}{6}$	$-\frac{\sqrt{3}}{6}$	$-\frac{\sqrt{3}}{6}$	$-\frac{1}{12}$	$\frac{1}{12}$	$-\frac{1}{12}$	$\frac{1}{12}$

By the help of the preceeding tables, one can determine the s.-a. MOs of every tetrahedral molecule arising from all types of atomic orbitals. Taking the 3jm symbols from Griffith [5], table C2.1., one calculates the coefficients $K(\gamma cp_c, Ai\varepsilon\varepsilon, ap_a)$ according to formula (5.2). In the paper [10] we already have elaborated the complete set of coefficients with respect to the equivalent set A (i.e. the P atoms of P_4 or P_4O_6, the H atoms of CH_4, and the ligands of many tetrehedral complexes). The most important AOs are those of species s/A_1 and p/T_2. Since the coefficients for the s-orbitals are trivially

$$K(cq, Aia, A_1 1) = \delta(c,a) \cdot (\vec{A}_i | Aaq),$$

we only repeat the coefficients for the p-orbitals in table 8. Because of different phases in the tables of Koster e.a.[21] and Griffith [5] we now get the opposite sign for the triads (ET_2T_2) and $(T_2T_2T_2)$.

Table 8. The coefficients $K(cq, Aia, T_2p)$

a c q	A_1 x	y^1	z	A_2 x	y^2	z	A_3 x	y^3	z	A_4 x	y^4	z
A_1T_2x	$\frac{1}{2}$	0	0	$\frac{1}{2}$	0	0	$\frac{1}{2}$	0	0	$\frac{1}{2}$	0	0
A_1T_2y	0	$\frac{1}{2}$	0	0	$\frac{1}{2}$	0	0	$\frac{1}{2}$	0	0	$\frac{1}{2}$	0
A_1T_2z	0	0	$\frac{1}{2}$	0	0	$\frac{1}{2}$	0	0	$\frac{1}{2}$	0	0	$\frac{1}{2}$
$T_2A_1 1$	$\frac{\sqrt{3}}{6}$	$\frac{\sqrt{3}}{6}$	$\frac{\sqrt{3}}{6}$	$\frac{\sqrt{3}}{6}$	$-\frac{\sqrt{3}}{6}$	$-\frac{\sqrt{3}}{6}$	$-\frac{\sqrt{3}}{6}$	$-\frac{\sqrt{3}}{6}$	$\frac{\sqrt{3}}{6}$	$-\frac{\sqrt{3}}{6}$	$\frac{\sqrt{3}}{6}$	$-\frac{\sqrt{3}}{6}$
$T_2E\ 1$	$\frac{\sqrt{6}}{12}$	$\frac{\sqrt{6}}{12}$	$-\frac{\sqrt{6}}{6}$	$\frac{\sqrt{6}}{12}$	$-\frac{\sqrt{6}}{12}$	$\frac{\sqrt{6}}{6}$	$-\frac{\sqrt{6}}{12}$	$-\frac{\sqrt{6}}{12}$	$-\frac{\sqrt{6}}{6}$	$-\frac{\sqrt{6}}{12}$	$\frac{\sqrt{6}}{12}$	$\frac{\sqrt{6}}{6}$
$T_2E\ 2$	$-\frac{1}{2}$	$\frac{1}{2}$	0	$-\frac{1}{2}$	$-\frac{1}{2}$	0	$\frac{1}{2}$	$-\frac{1}{2}$	0	$\frac{1}{2}$	$\frac{1}{2}$	0
$T_2T_1\xi$	0	$-\frac{\sqrt{2}}{4}$	$\frac{\sqrt{2}}{4}$	0	$\frac{\sqrt{2}}{4}$	$-\frac{\sqrt{2}}{4}$	0	$-\frac{\sqrt{2}}{4}$	$-\frac{\sqrt{2}}{4}$	0	$\frac{\sqrt{2}}{4}$	$\frac{\sqrt{2}}{4}$
$T_2T_1\eta$	$\frac{\sqrt{2}}{4}$	0	$-\frac{\sqrt{2}}{4}$	$-\frac{\sqrt{2}}{4}$	0	$-\frac{\sqrt{2}}{4}$	$\frac{\sqrt{2}}{4}$	0	$\frac{\sqrt{2}}{4}$	$-\frac{\sqrt{2}}{4}$	0	$\frac{\sqrt{2}}{4}$
$T_2T_1\zeta$	$-\frac{\sqrt{2}}{4}$	$\frac{\sqrt{2}}{4}$	0	$\frac{\sqrt{2}}{4}$	$\frac{\sqrt{2}}{4}$	0	$\frac{\sqrt{2}}{4}$	$-\frac{\sqrt{2}}{4}$	0	$-\frac{\sqrt{2}}{4}$	$-\frac{\sqrt{2}}{4}$	0
T_2T_2x	0	$-\frac{\sqrt{2}}{4}$	$-\frac{\sqrt{2}}{4}$	0	$\frac{\sqrt{2}}{4}$	$\frac{\sqrt{2}}{4}$	0	$-\frac{\sqrt{2}}{4}$	$\frac{\sqrt{2}}{4}$	0	$\frac{\sqrt{2}}{4}$	$-\frac{\sqrt{2}}{4}$
T_2T_2y	$-\frac{\sqrt{2}}{4}$	0	$-\frac{\sqrt{2}}{4}$	$\frac{\sqrt{2}}{4}$	0	$-\frac{\sqrt{2}}{4}$	$-\frac{\sqrt{2}}{4}$	0	$\frac{\sqrt{2}}{4}$	$\frac{\sqrt{2}}{4}$	0	$\frac{\sqrt{2}}{4}$
T_2T_2z	$-\frac{\sqrt{2}}{4}$	$-\frac{\sqrt{2}}{4}$	0	$\frac{\sqrt{2}}{4}$	$-\frac{\sqrt{2}}{4}$	0	$\frac{\sqrt{2}}{4}$	$\frac{\sqrt{2}}{4}$	0	$-\frac{\sqrt{2}}{4}$	$\frac{\sqrt{2}}{4}$	0

As a further example, we calculate the coefficients $K(cq, Bib, T_2p)$, which are needed for the MOs resulting from the p/T_2-orbitals of the oxygen atoms of P_4O_6. These coefficients are listed in table 9. The same coefficients apply to the symmetry coordinates of the molecule:

$$Q_q^{(Bb)c} = \sum_{ip} K(cq, Bib, T_2p) \cdot \Delta B_{ip}, \tag{22.30}$$

where $\Delta \vec{B}_i = (\Delta B_{ix}, \Delta B_{iy}, \Delta B_{iz})$ is the displacement vector of atom $\vec{B}_i$ [12].

Table 9. The coefficients $K(cq,Bib,T_2p)$

b c q \ p i	B_1			B_2			B_3			B_4			B_5			B_6		
	x	y^1	z	x	y^2	z	x	y^3	z	x	y^4	z	x	y^5	z	x	y^6	z
A_1T_2x	$\frac{\sqrt{6}}{6}$	0	0	$\frac{\sqrt{6}}{6}$	0	0	$\frac{\sqrt{6}}{6}$	0	0	$\frac{\sqrt{6}}{6}$	0	0	$\frac{\sqrt{6}}{6}$	0	0	$\frac{\sqrt{6}}{6}$	0	0
A_1T_2y	0	$\frac{\sqrt{6}}{6}$	0	0	$\frac{\sqrt{6}}{6}$	0	0	$\frac{\sqrt{6}}{6}$	0	0	$\frac{\sqrt{6}}{6}$	0	0	$\frac{\sqrt{6}}{6}$	0	0	$\frac{\sqrt{6}}{6}$	0
A_1T_2z	0	0	$\frac{\sqrt{6}}{6}$	0	0	$\frac{\sqrt{6}}{6}$	0	0	$\frac{\sqrt{6}}{6}$	0	0	$\frac{\sqrt{6}}{6}$	0	0	$\frac{\sqrt{6}}{6}$	0	0	$\frac{\sqrt{6}}{6}$
$E\,T_1\xi$	$\frac{1}{2}$	0	0	$-\frac{1}{2}$	0	0	0	0	0	$\frac{1}{2}$	0	0	$-\frac{1}{2}$	0	0	0	0	0
$E\,T_1\eta$	0	$-\frac{1}{2}$	0	0	0	0	0	$\frac{1}{2}$	0	0	$-\frac{1}{2}$	0	0	0	0	0	$\frac{1}{2}$	0
$E\,T_1\zeta$	0	0	0	0	0	$\frac{1}{2}$	0	0	$-\frac{1}{2}$	0	0	0	0	0	$\frac{1}{2}$	0	0	$-\frac{1}{2}$
$E\,T_2x$	$\frac{\sqrt{3}}{6}$	0	0	$\frac{\sqrt{3}}{6}$	0	0	$-\frac{\sqrt{3}}{3}$	0	0	$\frac{\sqrt{3}}{6}$	0	0	$\frac{\sqrt{3}}{6}$	0	0	$-\frac{\sqrt{3}}{3}$	0	0
$E\,T_2y$	0	$\frac{\sqrt{3}}{6}$	0	0	$-\frac{\sqrt{3}}{3}$	0	0	$\frac{\sqrt{3}}{6}$	0	0	$\frac{\sqrt{3}}{6}$	0	0	$-\frac{\sqrt{3}}{3}$	0	0	$\frac{\sqrt{3}}{6}$	0
$E\,T_2z$	0	0	$-\frac{\sqrt{3}}{3}$	0	0	$\frac{\sqrt{3}}{6}$	0	0	$\frac{\sqrt{3}}{6}$	0	0	$-\frac{\sqrt{3}}{3}$	0	0	$\frac{\sqrt{3}}{6}$	0	0	$\frac{\sqrt{3}}{6}$
$T_2A_1\,1$	0	0	$\frac{\sqrt{6}}{6}$	0	$\frac{\sqrt{6}}{6}$	0	$\frac{\sqrt{6}}{6}$	0	0	0	0	$\frac{-\sqrt{6}}{6}$	0	$\frac{-\sqrt{6}}{6}$	0	$\frac{-\sqrt{6}}{6}$	0	0
$T_2E\,1$	0	0	$-\frac{\sqrt{3}}{3}$	0	$\frac{\sqrt{3}}{6}$	0	$\frac{\sqrt{3}}{6}$	0	0	0	0	$\frac{\sqrt{3}}{3}$	0	$\frac{-\sqrt{3}}{6}$	0	$\frac{-\sqrt{3}}{6}$	0	0
$T_2E\,2$	0	0	0	0	$\frac{1}{2}$	0	$-\frac{1}{2}$	0	0	0	0	0	0	$-\frac{1}{2}$	0	$-\frac{1}{2}$	0	0
$T_2T_1\xi$	0	$\frac{1}{2}$	0	0	0	$-\frac{1}{2}$	0	0	0	0	$-\frac{1}{2}$	0	0	0	$\frac{1}{2}$	0	0	0
$T_2T_1\eta$	$-\frac{1}{2}$	0	0	0	0	0	0	0	$\frac{1}{2}$	$\frac{1}{2}$	0	0	0	0	0	0	0	$-\frac{1}{2}$
$T_2T_1\zeta$	0	0	0	$\frac{1}{2}$	0	0	0	$-\frac{1}{2}$	0	0	0	0	$-\frac{1}{2}$	0	0	0	$\frac{1}{2}$	0
T_2T_2x	0	$-\frac{1}{2}$	0	0	0	$-\frac{1}{2}$	0	0	0	0	$\frac{1}{2}$	0	0	0	$\frac{1}{2}$	0	0	0
T_2T_2y	$-\frac{1}{2}$	0	0	0	0	0	0	0	$-\frac{1}{2}$	$\frac{1}{2}$	0	0	0	0	0	0	0	$\frac{1}{2}$
T_2T_2z	0	0	0	$-\frac{1}{2}$	0	0	0	$-\frac{1}{2}$	0	0	0	0	$\frac{1}{2}$	0	0	0	$\frac{1}{2}$	0

22.3. Polyhedral isoscalar factors

Whereas the elaboration of the s.-a. linear combinations of orbitals and coordinates is a conventional technique, we now come to the central point of our innovation, the group theoretical description of the topological structures by the various polyhedral isoscalar factors.

The four atoms of P_4 occupy the positions $\vec{A}_i$ given in table 2. The atomic orbitals at these centres define two types of two-centre integrals distinguished by the edge vectors connecting the centres.

1) Both orbitals are located at the same centre, i.e. the edge vector degenerates to $\vec{0}_i^A = \vec{A}_i - \vec{A}_i \sim \vec{A}_i$, cf. page 21. Here and in the following,

the equivalence of sets, i.e. $\vec{M}_i \sim \vec{N}_i$ for instance, implies such a nume-
ration that $\sigma_{ik}^M(g) = \sigma_{ik}^N(g)$ in addition to the ordinary equivalence of
the induced representations, $\sigma^M \sim \sigma^N$. But this does not imply $\vec{M}_i = \vec{N}_i$! As
shown in section 3, this equivalence, $\vec{M}_i \sim \vec{N}_i$, leads to $(\vec{M}_i | M\mu mp) =$
$(\vec{N}_i | N\mu mp)$. This means in the present case: $(\vec{O}_i^A | O^A \alpha a p_a) = (\vec{A}_i | A\alpha a p_a)$.
Since $\vec{O}_i^A$ belongs to position $\vec{A}_i$ only, the topological matrix (3.1) is
given by:

$$\tau\left(\begin{smallmatrix} -AAO^A \\ ikl \end{smallmatrix}\right) = \delta(i,k)\delta(k,l)/\sqrt{Z(A)} \qquad (22.31)$$

2) The orbitals are located at different centres. In contrast to set B,
there is only one type of coordination within the set A indicated by
the edge vectors $\vec{S}_{ik}^A = \vec{A}_i - \vec{A}_k$. These twelve vectors lie in the reflection
planes and therefore we have the equivalence $\sigma^{SA} \sim \sigma^C$. An adequate enumeration of the edges then allows $\vec{S}_k^A \sim \vec{C}_k$. This correspondence can be
made explicit by an appropriate choice of the coefficients in the rela-
tion (3.25): $\vec{S}_{1k}^{A\prime} = \mu_1 \vec{A}_i + \mu_2 \vec{A}_k$. If we take $\mu_1 = (1+\sqrt{2})\sqrt{3}/4$ and $\mu_2 = (-1+\sqrt{2})\sqrt{3}/4$,
we get $\vec{S}_{12}^{A\prime} = \vec{C}_1$ etc. We now define the topological matrix of the trian-
gles $-AAS^A$ by:

$$\tau\left(\begin{smallmatrix} -AAS^A \\ ikl \end{smallmatrix}\right) = \delta(\mu_1\vec{A}_i + \mu_2\vec{A}_k, \vec{C}_1)/\sqrt{Z(C)} , \qquad (22.32)$$

or equivalently

$$\sum_{i \neq k}\tau\left(\begin{smallmatrix} -AAS^A \\ ikl \end{smallmatrix}\right)\cdot(\mu_1\vec{A}_i + \mu_2\vec{A}_k)\sqrt{Z(-AAS^A)} = \vec{C}_1 \qquad (22.33)$$

The topological correlations expressed by (22.31 and 32) are listed in
table 10.

Table 10. Non-zero matrix elements of $\tau\left(\begin{smallmatrix} -AAO^A \\ ikl \end{smallmatrix}\right)$ and $\tau\left(\begin{smallmatrix} -AAS^A \\ ikl \end{smallmatrix}\right)$

A_i	1 2 3 4	A_i	1	1	1	2	2	2	3	3	3	4	4	4
A_k	1 2 3 4	A_k	2	3	4	1	3	4	1	2	4	1	2	3
$O_1^A \sim A_1$	1 2 3 4	$S_1^A \sim C_1$	1	3	2	4	11	9	6	8	10	5	12	7

Basing upon these correlations, we calculate the polyhedral isoscalar
factors of the triangles $-AAO^A$ and $-AAS^A$ according to formula (6.6).
The result is compiled in table 11 on the following page. It differs
from that of [11] because of the different SALC coefficients of table 6
and the different order of the triple product $a \times b \times c$.

In the same way, we treat the equivalent set B. The six centres listed
in table 2 are occupied by the oxygen atoms of P_4O_6. There are now three
types of coordination within this set:
1) The orbitals are located at the same centre again, i.e. the edge vec-
tors are $\vec{O}_k^B = \vec{B}_k - \vec{B}_k \sim \vec{B}_k$. The topological correspondence is quite analogous to (22.31):

$$\tau\left(\begin{smallmatrix} -BBO^B \\ ikl \end{smallmatrix}\right) = \delta(i,k)\delta(k,l)/\sqrt{Z(B)} \qquad (22.34)$$

2) The orbitals are located at adjoining centres (for instance $\vec{B}_1$ and

Table 11. Polyhedral isoscalars $\mathrm{PIs}\!\left(\begin{smallmatrix}-AAO^A\\abc\end{smallmatrix}\right)$ and $\mathrm{PIs}\!\left(\begin{smallmatrix}-AAS^A\\ \gamma \\ abc\end{smallmatrix}\right)$

a b γc	$-AAO^A$(alg.)	$-AAO^A$(num.)	$-AAS^A$(alg.)	$-AAS^A$(num.)
$A_1A_1\ A_1$	$1/4$	0.25000000	$1/4$	0.25000000
$T_2T_2\ A_1$	$\sqrt{3}/4$	0.43301270	$-\sqrt{3}/12$	-0.14433757
$T_2T_2\ E$	-	-	$-1/\sqrt{6}$	-0.40824829
$T_2T_2\ T_1$	-	-	$1/2$	0.50000000
$A_1T_21T_2$	$\sqrt{3}/4$	0.43301270	$(\sqrt{2}-2)/8$	-0.07322330
$T_2A_11T_2$	$\sqrt{3}/4$	0.43301270	$(\sqrt{2}+2)/8$	0.42677670
$T_2T_21T_2$	$-\sqrt{6}/4$	-0.61237244	$1/4$	0.25000000
$A_1T_22T_2$	-	-	$-(\sqrt{2}+2)/8$	-0.42677670
$T_2A_12T_2$	-	-	$-(\sqrt{2}-2)/8$	0.07322330
$T_2T_22T_2$	-	-	$-1/4$	-0.25000000

$\vec{B}_2$). The 24 edge vectors are termed $\vec{S}_1^B$ and are equivalent to the set D: $\vec{S}_1^B \sim \vec{D}_1$. The description of this topological correspondence is complicated by the fact that the vectors $\mu_1\vec{B}_i+\mu_2\vec{B}_k$ having one zero component may be equivalent but not equal to a vector of set D. In order to produce an identity, we resort to a certain vector product, which is well defined with respect to the group T_d:

$$[\vec{r}\times\vec{r}]_+ = (yz'+zy')\vec{e}_1 + (zx'+xz')\vec{e}_2 + (xy'+yx')\vec{e}_3 \qquad (22.35)$$

The definition

$$\vec{S}_{ik}^B = (\sqrt{3}/3)\vec{B}_i + (\sqrt{5}/3)\vec{B}_k + (1/3)[\vec{B}_i\times\vec{B}_k]_+ \qquad (22.36)$$

then yields $\vec{S}_{11}^B=\vec{D}_1$ etc. The full list of the correlations is given in table 12, which also indicates the non-zero matrix elements of the topological matrix:

$$\tau\!\left(\begin{smallmatrix}-BBS^B\\ikl\end{smallmatrix}\right) = \delta(\sqrt{3}\vec{B}_i+\sqrt{5}\vec{B}_k+[\vec{B}_i\times\vec{B}_k]_+,3\vec{D}_1)/\sqrt{Z(D)} \qquad (22.37)$$

An equivalent expression is:

$$\sum_{ik}\tau\!\left(\begin{smallmatrix}-BBS^B\\ikl\end{smallmatrix}\right)\sqrt{Z(-BBS^B)}\cdot(\sqrt{\tfrac{3}{3}}\vec{B}_i+\sqrt{\tfrac{5}{3}}\vec{B}_k+\tfrac{1}{3}[\vec{B}_i\times\vec{B}_k]_+) = \vec{D}_1 \qquad (22.38)$$

3) The third possibility is the location of the orbitals at the opposite positions like $\vec{B}_1$ and $\vec{B}_4$. The six edge vectors of this type are termed $\vec{T}_k^B \sim \vec{B}_k$ and the topological matrix is in this case:

$$\tau\!\left(\begin{smallmatrix}-BBT^B\\ikl\end{smallmatrix}\right) = \delta(\vec{B}_i-\vec{B}_k,2\vec{B}_1)/\sqrt{Z(B)} \qquad (22.39)$$

The essence of (22.34/37/39) is gathered in table 12. This compilation then allows the calculation of the isoscalar factors of the triangles of the type $-BBO^B$, $-BBS^B$, and $-BBT^B$. These are listed in the tables 13 and 14.

Table 12. Non-zero matrix elements of $\tau(-\overline{BBO}^B_{ikl})$, $\tau(-\overline{BBT}^B_{ikl})$, and $\tau(-\overline{BBS}^B_{ikl})$

B_i	1	2	3	4	5	6	B_i	1	2	3	4	5	6	B_i	1	1	1	1
B_k	1	2	3	4	5	6	B_k	4	5	6	1	2	3	B_k	2	3	5	6
$O^B_1 \sim B_1$	1	2	3	4	5	6	$T^B_1 \sim B_1$	1	2	3	4	5	6	$S^B_1 \sim D_1$	1	15	4	14

B_i	2	2	2	2	3	3	3	3	4	4	4	4	5	5	5	5	6	6	6	6
B_k	1	3	4	6	1	2	4	5	2	3	5	6	1	3	4	6	1	2	4	5
$S^B_1 \sim D_1$	20	5	19	6	9	22	11	24	3	16	2	13	18	8	17	7	10	23	12	21

Table 13. Polyhedral isoscalars $PIs(-\overline{BBO}^B_{abc})$ and $PIs(-\overline{BBT}^B_{abc})$

a b c	$-\overline{BBO}^B$(alg.)	$-\overline{BBO}^B$(num.)	$-\overline{BBT}^B$(alg.)	$-\overline{BBT}^B$(num.)
$A_1A_1A_1$	$1/6$	0.16666667	$1/6$	0.16666667
$E\,E\,A_1$	$\sqrt{2}/6$	0.23570226	$\sqrt{2}/6$	0.23570226
$T_2T_2A_1$	$\sqrt{3}/6$	0.28867513	$-\sqrt{3}/6$	-0.28867513
$A_1E\,E$	$\sqrt{2}/6$	0.23570226	$\sqrt{2}/6$	0.23570226
$E\,A_1E$	$\sqrt{2}/6$	0.23570226	$\sqrt{2}/6$	0.23570226
$E\,E\,E$	$-\sqrt{2}/6$	-0.23570226	$-\sqrt{2}/6$	-0.23570226
T_2T_2E	$-1/\sqrt{6}$	-0.40824829	$1/\sqrt{6}$	0.40824829
$A_1T_2T_2$	$\sqrt{3}/6$	0.28867513	$-\sqrt{3}/6$	-0.28867513
$T_2A_1T_2$	$\sqrt{3}/6$	0.28867513	$\sqrt{3}/6$	0.28867513
$E\,T_2T_2$	$-1/\sqrt{6}$	-0.40824829	$1/\sqrt{6}$	0.40824829
$T_2E\,T_2$	$-1/\sqrt{6}$	-0.40824829	$-1/\sqrt{6}$	-0.40824829
$T_2T_2T_2$	0	0.00000000	0	0.00000000

Table 14. Polyhedral isoscalars $PIs\begin{pmatrix} -\overline{BBS}^B \\ \gamma \\ abc \end{pmatrix}$

a b γc	algebraic	numerical	a b γc	algebraic	numerical
$A_1A_1\,A_1$	$1/6$	0.16666667	$E\,T_2 1T_1$	$\sqrt{3}/6$	0.28867513
$E\,E\,A_1$	$-\sqrt{2}/12$	-0.11785113	$T_2E\,1T_1$	$\sqrt{5}/12$	0.18633900
$T_2T_2\,A_1$	0	0.00000000	$T_2T_2 1T_1$	$-1/12$	-0.08333333
$E\,E\,A_2$	$-\sqrt{6}/12$	-0.20412414	$E\,T_2 2T_1$	0	0.00000000
$A_1E\,1E$	0	0.00000000	$T_2E\,2T_1$	$-\sqrt{3}/12$	-0.14433757
$E\,A_1 1E$	$\sqrt{6}/12$	0.20412414	$T_2T_2 2T_1$	$-\sqrt{15}/12$	-0.32274861
$E\,E\,1E$	$\sqrt{6}/12$	0.20412414	$E\,T_2 3T_1$	$-\sqrt{6}/12$	-0.20412414
$T_2T_2 1E$	0	0.00000000	$T_2E\,3T_1$	$\sqrt{10}/12$	0.26352314
$A_1E\,2E$	$-\sqrt{2}/6$	-0.23570226	$T_2T_2 3T_1$	$-\sqrt{2}/12$	-0.11785113
$E\,A_1 2E$	$\sqrt{2}/12$	0.11785113	$A_1T_2 1T_2$	$1/6$	0.16666667
$E\,E\,2E$	$-\sqrt{2}/12$	-0.11785113	$T_2A_1 1T_2$	$\sqrt{15}/18$	0.21516574
$T_2T_2 2E$	0	0.00000000	$E\,T_2 1T_2$	$\sqrt{2}/12$	0.11785113

Table 14. (continued)

a b γc	algebraic	numerical	a b γc	algebraic	numerical
$T_2E\ 1T_2$	$\sqrt{30}/36$	0.15214515	$T_2T_22T_2$	$-\sqrt{15}/12$	-0.32274861
$T_2T_21T_2$	$-\sqrt{2}/12$	-0.11785113	$A_1T_23T_2$	$-\sqrt{2}/6$	-0.23570226
$A_1T_22T_2$	0	0.00000000	$T_2A_13T_2$	$\sqrt{30}/36$	0.15214515
$T_2A_12T_2$	$-\sqrt{2}/12$	-0.11785115	$E\ T_23T_2$	$-1/6$	-0.16666667
$E\ T_22T_2$	0	0.00000000	$T_2E\ 3T_2$	$\sqrt{15}/36$	0.10758287
$T_2E\ 2T_2$	$-1/12$	-0.08333333	$T_2T_23T_2$	$-1/12$	-0.08333333

In the molecule P_4O_6, there are further triangles (or two-centre integrals) involving one position of set A and one of set B. The edge vectors between directly adjoining centres are termed $\vec{S}_{ik}^{AB}=\vec{A}_i-\vec{B}_k$ (for instance $\vec{B}_1$ and $\vec{A}_1$), whereas the indirect connection (for instance between $\vec{A}_1$ and $\vec{B}_4$) is expressed by $\vec{T}_{ik}^{AB}=\vec{A}_i-\vec{B}_k$. There are twelve vectors in each set and thus $\vec{S}_1^{AB}\sim\vec{T}_1^{AB}\sim\vec{C}_1$. The topological matrices are given by:

$$\tau\left(\begin{smallmatrix}-\mathrm{ABS}^{AB}\\ikl\end{smallmatrix}\right)=\delta(\sqrt{3}\vec{A}_i+(\sqrt{2}-1)\vec{B}_k,2\vec{C}_1)/\sqrt{Z(C)} \qquad (22.40)$$

$$\tau\left(\begin{smallmatrix}-\mathrm{ABT}^{AB}\\ikl\end{smallmatrix}\right)=\delta(\sqrt{3}\vec{A}_i-(\sqrt{2}-1)\vec{B}_k,2\vec{C}_1)/\sqrt{Z(C)} \qquad (22.41)$$

The correspondences defined in this way are compiled in table 15.

Table 15. Non-zero matrix elements of $\tau\left(\begin{smallmatrix}-\mathrm{ABS}^{AB}\\ikl\end{smallmatrix}\right)$ and $\tau\left(\begin{smallmatrix}-\mathrm{ABT}^{AB}\\ikl\end{smallmatrix}\right)$

A_i	1	1	1	1	1	1	2	2	2	2	2	2	3	3	3	3	3	3	4	4	4	4	4	4
B_k	1	2	3	4	5	6	1	2	3	4	5	6	1	2	3	4	5	6	1	2	3	4	5	6
$S_1^{AB}\sim C_1$	3	2	1	-	-	-	-	-	4	9	11	-	6	-	-	-	8	10	-	5	-	12	-	7
$T_1^{AB}\sim C_1$	-	-	-	3	2	1	9	11	-	-	-	4	-	8	10	6	-	-	12	-	7	-	5	-

Finally, the polyhedral isoscalar factors calculated from these correlations are given in table 16.

Table 16. Polyhedral isoscalars $\mathrm{PIs}\left(\begin{smallmatrix}-\mathrm{ABS}^{AB}\\ \gamma\\ abc\end{smallmatrix}\right)$ and $\mathrm{PIs}\left(\begin{smallmatrix}-\mathrm{ABT}^{AB}\\ \gamma\\ abc\end{smallmatrix}\right)$

a b γc	$-\mathrm{ABS}^{AB}$(alg.)	$-\mathrm{ABS}^{AB}$(num.)	$-\mathrm{ABT}^{AB}$(alg.)	$-\mathrm{ABT}^{AB}$(num.)
$A_1A_1\ A_1$	$\sqrt{6}/12$	0.20412414	$\sqrt{6}/12$	0.20412414
$T_2T_2\ A_1$	$\sqrt{6}/12$	0.20412414	$-\sqrt{6}/12$	-0.20412414
$A_1E\ E$	$\sqrt{3}/6$	0.28867513	$\sqrt{3}/6$	0.28867513
$T_2T_2\ E$	$-\sqrt{3}/6$	-0.28867513	$\sqrt{3}/6$	0.28867513
$T_2E\ T_1$	$\sqrt{2}/4$	0.35355339	$\sqrt{2}/4$	0.35355339
$T_2T_2\ T_1$	$\sqrt{2}/4$	0.35355339	$-\sqrt{2}/4$	-0.35355339
$A_1T_21T_2$	$1/4$	0.25000000	$-1/4$	-0.25000000
$T_2A_11T_2$	$(\sqrt{6}+\sqrt{3})/12$	0.34846171	$(\sqrt{6}+\sqrt{3})/12$	0.34846171
$T_2E\ 1T_2$	$(\sqrt{3}-\sqrt{6})/12$	-0.05978658	$(\sqrt{3}-\sqrt{6})/12$	-0.05978658

Table 16. (continued)

a b γc	$-ABS^{AB}$(alg.)	$-ABS^{AB}$(num.)	$-ABT^{AB}$(alg.)	$-ABT^{AB}$(num.)
$T_2T_21T_2$	$-1/4$	-0.25000000	$1/4$	0.25000000
$A_1T_22T_2$	$-1/4$	-0.25000000	$1/4$	0.25000000
$T_2A_12T_2$	$(\sqrt{6}-\sqrt{3})/12$	0.05978658	$(\sqrt{6}-\sqrt{3})/12$	0.05978658
$T_2E\ 2T_2$	$(\sqrt{3}+\sqrt{6})/12$	0.34846171	$(\sqrt{3}+\sqrt{6})/12$	0.34846171
$T_2T_22T_2$	$-1/4$	-0.25000000	$1/4$	0.25000000

In contrast to the edge vectors connecting centres of the same set, S^A for instance, we have to consider S^{AB} and the set of the inverted vectors S^{BA} as sets of inequivalent objects. But both set induce the same representation σ^C and an appropriate numbering yields $\vec{S}_1^{BA} \sim \vec{S}_1^{AB} \sim \vec{C}_1$. The same applies to $\vec{T}_1^{BA} \sim \vec{T}_1^{AB} \sim \vec{C}_1$. We therefore may choose

$$\tau\binom{-BAS^{BA}}{ikl} = \tau\binom{-ABS^{AB}}{kil} \text{ and } \tau\binom{-BAT^{BA}}{ikl} = \tau\binom{-ABT^{AB}}{kil}, \tag{22.42}$$

which yiels simple relations of the polyhedral isoscalar factors:

$$PIs\begin{pmatrix} -BAS^{BA} \\ \gamma \\ bac \end{pmatrix} = \{abc\}\cdot PIs\begin{pmatrix} -ABS^{AB} \\ \gamma \\ abc \end{pmatrix}, \quad PIs\begin{pmatrix} -BAT^{BA} \\ \gamma \\ bac \end{pmatrix} = \{abc\}\cdot PIs\begin{pmatrix} -BAT^{BA} \\ \gamma \\ bac \end{pmatrix} \tag{22.43}$$

At this point one may ask, whether the choice of the topological matrices is unequivocal. Indeed, the matrices are not determined unambiguously, but a different choice of the topological correlations causes only a unitary transformation of the polyhedral isoscalars (a change of phase in the multiplicity free cases). We demonstrate this by an example. If we replace (22.40) by

$$\tau'\binom{-ABS^{AB}}{ikl} = \delta(-\sqrt{3}\vec{A}_1 + (\sqrt{2}+1)\vec{B}_k, 2\vec{C}_1)/\sqrt{Z(C)}, \tag{22.44}$$

the first correlated triple, for instance, is $\vec{A}_1, \vec{B}_1, \vec{C}_6$ instead of $\vec{A}_1$, $\vec{B}_1, \vec{C}_3$ and we must rearrange table 15. The resulting new polyhedral isoscalars are related to those of table 16 by the transformation

$$PIs\begin{pmatrix} -ABS^{AB} \\ \gamma \\ abc \end{pmatrix} = \sum_\sigma u(c)_{\gamma\sigma} PIs\begin{pmatrix} -ABS^{AB} \\ \sigma \\ abc \end{pmatrix} \tag{22.45}$$

with $u(A_1)=+1$, $u(E)=+1$, $u(T_1)=-1$, $u(T_2)_{11}=u(T_2)_{22}=0$, $u(T_2)_{12}=u(T_2)_{21}=-1$.

Similar relations, caused by different correlations between the same sets, are found by inspecting the tables 13 and 16:

$$PIs\binom{-BBO^B}{abc} = \varphi(b)\cdot PIs\binom{-BBT^B}{abc}, \quad PIs\begin{pmatrix} -ABS^{AB} \\ \gamma \\ abc \end{pmatrix} = \varphi(b)\cdot PIs\begin{pmatrix} -ABT^{AB} \\ \gamma \\ abc \end{pmatrix} \tag{22.46}$$

with the phase factors $\varphi(A_1)=+1$, $\varphi(E)=+1$, and $\varphi(T_2)=-1$.

The factors calculated so far also apply to strucures like Be_4Cl_4 and $[Cu(CN)_4]^{2-}$, where the positions of type A occur twice, let us say A and A'. The isoscalars within the set A' are trivially equal to those of set A. There are two types of edge vectors connecting centres of

both sets. If both centres lie on the same axis, we have $\vec{U}_i^{AA'} = \vec{A}_i - \vec{A}_i' \sim \vec{A}_i$.
In the other case the distance vectors $\vec{V}^{AA'}$ induce the representation σ^C
and therefore $\vec{V}_1^{AA'} \sim \vec{S}_1^A \sim \vec{C}_1$. From these equivalences follows:

$$PIs\left(\begin{smallmatrix}-A\hat{A}U^{AA'}\\ikl\end{smallmatrix}\right) = PIs\left(\begin{smallmatrix}-AAO^A\\ikl\end{smallmatrix}\right), \quad PIs\left(\begin{matrix}-AA'V^{AA'}\\ \gamma\\ abc\end{matrix}\right) = PIs\left(\begin{matrix}-AAS^A\\ \gamma\\ abc\end{matrix}\right) \tag{22.47}$$

The molecule $C(CH_3)_4$ requires the additional calculation of the polyhedral isoscalars involving the positions of set C.

22.4. Polyhedral isoscalar factors of the second kind

The next step is the consideration of the triangles subtended by
three atomic centres. Within the scope of this principal case study, we
calculate the complete set of polyhedral isoscalars of the second kind
involving three centres of type A. With regard to the main application
of these isoscalars in (8.14), one can confine the calculation to the
subclass (8.15). In set A, there are five different types of triangles
corresponding to possible three-centre integrals:
1) All three centres coincide. We mark these "null triangles" by Δ_o^A.
There are naturally four triangles of this type, i.e. $\Delta_{oi}^A \sim \vec{A}_i$. The topological matrices of the second kind according to (7.16) correlate each
triangle with its second vertex $\vec{A}_i$ and with the edge vector connecting
the first and third vertex, i.e. with $\vec{O}_i^A$:

$$\tau^2\left(\begin{smallmatrix}\Delta^A & O^A & A\\ i & k & l\end{smallmatrix}\right) = \tau\left(\begin{smallmatrix}-AAO^A\\ikl\end{smallmatrix}\right) = \delta(i,k)\delta(k,l)/\sqrt{Z(A)} \tag{22.48}$$

2) The first and the third centre coincide. These degenerate triangles
Σ_o^A correspond to the edge vectors connecting the first and second centre, i.e. $\Sigma_{ol}^A \sim \vec{S}_1^A \sim \vec{C}_1$ and

$$\tau^2\left(\begin{smallmatrix}\Sigma^A & O^A & A\\ i & k & l\end{smallmatrix}\right) = \tau\left(\begin{smallmatrix}-AAS^A\\kli\end{smallmatrix}\right). \tag{22.49}$$

3) The first and the second centre coincide. This set is termed Σ_1^A:
$\Sigma_{11}^A \sim \vec{S}_1^A \sim \vec{C}_1$ again. In this case, it is easier to express the correlations
in a first step by the topological matrix (7.7) and to calculate τ^2 by
the inversion of (7.17):

$$\tau^2\left(\begin{smallmatrix}\Delta & SC\\irm\end{smallmatrix}\right) = \sum_{kl}\tau\left(\begin{smallmatrix}\Delta & ACB\\ikml\end{smallmatrix}\right)\tau\left(\begin{smallmatrix}-ABS\\klr\end{smallmatrix}\right)\sqrt{Z(-ABS)} \tag{22.50}$$

In the present case we have

$$\tau\left(\begin{smallmatrix}\Sigma^A_1 AAA\\1 1 kml\end{smallmatrix}\right) = \tau\left(\begin{smallmatrix}-AAS^A\\kli\end{smallmatrix}\right)\delta(k,m) \tag{22.51}$$

and (22.50) yields:

$$\tau^2\left(\begin{smallmatrix}\Sigma^A_1 S^A A\\1 1 r m\end{smallmatrix}\right) = \sum_l \tau\left(\begin{smallmatrix}-AAS^A\\mli\end{smallmatrix}\right)\tau\left(\begin{smallmatrix}-AAS^A\\mlr\end{smallmatrix}\right)\sqrt{Z(-AAS^A)} \tag{22.51}$$

The non-zero matrix elements taken from table 10 are given in table 17.
4) The second and the third centre coincide. This set Σ_2^A is different,
but equivalent to Σ_1^A, i.e. $\Sigma_{21}^A \sim \vec{S}_1^A \sim \vec{C}_1$. In this case, the correlations are
given by

$$\tau\left(\begin{smallmatrix}\Sigma_1^A 2AAA\\ {}_1 \text{kml}\end{smallmatrix}\right) = \delta(m,1)\tau\left(\begin{smallmatrix}-AAS^A\\ \text{kmi}\end{smallmatrix}\right) \tag{22.53}$$

and via (22.50) by

$$\tau^2\left(\begin{smallmatrix}\Sigma_2^A S^A A\\ {}_1 2_r \text{ m}\end{smallmatrix}\right) = \sum_k \tau\left(\begin{smallmatrix}-AAS^A\\ \text{kmi}\end{smallmatrix}\right)\tau\left(\begin{smallmatrix}-AAS^A\\ \text{kmr}\end{smallmatrix}\right)\sqrt{Z(-AAS^A)} \tag{22.54}$$

The result is again compiled in table 17.

5) Finally there is one set of 24 ordinary triangles termed Δ^A with $\Delta_m^A \sim \vec{D}_m$. In the sense of (7.8), the correlation of the triangle (i.e. the number of D) and the vertex numbers is achieved by the linear combination of the position vectors $\mu'_1\vec{A}_k + \mu'_2\vec{A}_m + \mu'_3\vec{A}_1$ with $\mu'_1 = (1+\sqrt{5})/2\sqrt{3}$, $\mu'_2 = (1-\sqrt{3})/2\sqrt{3}$, and $\mu'_3 = (\sqrt{5}-\sqrt{3})/2\sqrt{3}$. We therefore define the topological matrix by

$$\tau\left(\begin{smallmatrix}\Delta^A AAA\\ {}_i \text{ kml}\end{smallmatrix}\right) = \delta(\mu'_1\vec{A}_k + \mu'_2\vec{A}_m + \mu'_3\vec{A}_1, \vec{D}_i)/\sqrt{Z(D)} \tag{22.55}$$

With (22.32) eq.(22.50) results in:

$$\tau^2\left(\begin{smallmatrix}\Delta^A S^A A\\ {}_i \text{ r m}\end{smallmatrix}\right) = \sum_{kl}\delta(\mu'_1\vec{A}_k + \mu'_2\vec{A}_m + \mu'_3\vec{A}_1, \vec{D}_i)\delta(\mu_1\vec{A}_k + \mu_2\vec{A}_1, \vec{C}_r)/\sqrt{Z(D)} \tag{22.56}$$

These correlations are listed in table 18.

Table 17. Non-zero matrix elements of $\tau^2\left(\begin{smallmatrix}\Sigma_1^A S^A A\\ {}_1_r \text{ m}\end{smallmatrix}\right)$ and $\tau^2\left(\begin{smallmatrix}\Sigma_2^A S^A A\\ {}_1_r \text{ m}\end{smallmatrix}\right)$

A_m	1	1	1	1	1	1	2	2	2	2	2	2	3	3	3	3	3	3	4	4	4	4	4	4
$S_r^A \sim C_r$	1	2	3	4	5	6	1	4	8	9	11	12	3	6	7	8	10	11	2	5	7	9	10	12
$\Sigma_{11}^A \sim C_1$	1	2	3	-	-	-	-	4	-	9	11	-	-	6	-	8	10	-	-	5	7	-	-	12
$\Sigma_{21}^A \sim C_1$	-	-	-	1	2	3	4	-	11	-	-	9	6	-	10	-	-	8	5	-	-	12	7	-

Table 18. Non-zero matrix elements of $\tau^2\left(\begin{smallmatrix}\Delta^A S^A A\\ {}_i \text{ r m}\end{smallmatrix}\right)$

A_m	1	1	1	1	1	1	2	2	2	2	2	2	3	3	3	3	3	3	4	4	4	4	4	4
$S_r^A \sim C_r$	8	11	12	9	7	10	3	6	2	5	7	4	1	9	12	2	5	4	1	8	11	6	3	
$\Delta_i^A \sim D_i$	11	17	16	2	8	24	1	14	20	10	23	7	22	13	3	9	18	21	5	19	12	4	15	

From (7.22) now follow the polyhedral isoscalar factors of the second kind. Because of (22.48 and 49), the factors of the triangles Δ_o^A and Σ_o^A are very simple:

$$PIs^2\left(\begin{smallmatrix}\Delta_o^A O^A A\\ \text{a b c}\end{smallmatrix}\right) = PIs\left(\begin{smallmatrix}-A & A & O^A\\ \text{a} & \text{b} & \text{c}\end{smallmatrix}\right) \tag{22.57}$$

$$PIs^2\left(\begin{smallmatrix}\Sigma_o^A O^A A\\ \alpha\\ \text{a b c}\end{smallmatrix}\right) = PIs\left(\begin{smallmatrix}-A & A & S^A\\ \alpha\\ \text{b} & \text{c} & \text{a}\end{smallmatrix}\right) \tag{22.58}$$

There remains the tabulation for the triangles Σ_1^A, Σ_2^A, and Δ^A. It is given in the tables 19 and 20. Comparing the entries of table 17, we get the following relation, where the unitary matrix is that of (22.45):

$$PIs^2\left(\begin{smallmatrix}\Sigma_2^A S^A A\\ \alpha^2\beta\\ \text{a b c}\end{smallmatrix}\right) = \sum_{\beta'} u(b)_{\beta\beta'}PIs^2\left(\begin{smallmatrix}\Sigma_1^A S^A A\\ \alpha^1\beta\\ \text{a b c}\end{smallmatrix}\right) \tag{22.59}$$

Table 19. Polyhedral isoscalars $\text{PIs}^2\begin{pmatrix}\Sigma_1^A S^A A \\ \alpha^1\beta \\ a\ b\ c\end{pmatrix}$ and $\text{PIs}^2\begin{pmatrix}\Sigma_2^A S^A A \\ \alpha^2\beta \\ a\ b\ c\end{pmatrix}$

$\alpha a\ \beta b\ c$	$\Sigma_1^A S^A A$(alg.)	$\Sigma_1^A S^A A$(num.)	$\Sigma_2^A S^A A$(alg.)	$\Sigma_2^A S^A A$(num.)
$A_1\ A_1A_1$	$\sqrt{3}/12$	0.14433757	$\sqrt{3}/12$	0.14433757
$A_1\ 1T_2T_2$	$(2\sqrt{3}+\sqrt{6})/24$	0.24639964	$(\sqrt{6}-2\sqrt{3})/24$	-0.04227549
$A_1\ 2T_2T_2$	$(2\sqrt{3}-\sqrt{6})/24$	0.04227549	$-(\sqrt{6}+2\sqrt{3})/24$	-0.24639964
$E\ E\ A_1$	$\sqrt{6}/12$	0.20412415	$\sqrt{6}/12$	0.20412415
$E\ T_1T_2$	$1/4$	0.25000000	$-1/4$	-0.25000000
$E\ 1T_2T_2$	$(\sqrt{6}-2\sqrt{3})/24$	-0.04227549	$-(2\sqrt{3}+\sqrt{6})/24$	-0.24639964
$E\ 2T_2T_2$	$(\sqrt{6}+2\sqrt{3})/24$	0.24639964	$(2\sqrt{3}-\sqrt{6})/24$	0.04227549
$T_1\ T_1A_1$	$1/4$	0.25000000	$-1/4$	-0.25000000
$T_1\ E\ T_2$	$-1/4$	-0.25000000	$-1/4$	-0.25000000
$T_1\ T_1T_2$	$\sqrt{2}/8$	0.17677670	$-\sqrt{2}/8$	-0.17677670
$T_1\ 1T_2T_2$	$(1-\sqrt{2})/8$	-0.05177670	$-(1+\sqrt{2})/8$	-0.30177670
$T_1\ 2T_2T_2$	$(1+\sqrt{2})/8$	0.30177670	$-(1-\sqrt{2})/8$	0.05177670
$1T_2\ 1T_2A_1$	$1/4$	0.25000000	0	0.00000000
$1T_2\ 2T_2A_1$	0	0.00000000	$-1/4$	-0.25000000
$1T_2\ A_1T_2$	$(\sqrt{6}+2\sqrt{3})/24$	0.24639964	$(\sqrt{6}+2\sqrt{3})/24$	0.24639964
$1T_2\ E\ T_2$	$(\sqrt{6}-2\sqrt{3})/24$	-0.04227549	$(\sqrt{6}-2\sqrt{3})/24$	-0.04227549
$1T_2\ T_1T_2$	$(1-\sqrt{2})/8$	-0.05177670	$-(1-\sqrt{2})/8$	0.05177670
$1T_2\ 1T_2T_2$	$-(4+\sqrt{2})/16$	0.33838835	$\sqrt{2}/16$	0.08838835
$1T_2\ 2T_2T_2$	$-\sqrt{2}/16$	-0.08838835	$(4+\sqrt{2})/16$	0.33838835
$2T_2\ 1T_2A_1$	0	0.00000000	$-1/4$	-0.25000000
$2T_2\ 2T_2A_1$	$1/4$	0.25000000	0	0.00000000
$2T_2\ A_1T_2$	$(2\sqrt{3}-\sqrt{6})/24$	0.04227549	$(2\sqrt{3}-\sqrt{6})/24$	0.04227549
$2T_2\ E\ T_2$	$(2\sqrt{3}+\sqrt{6})/24$	0.24639964	$(2\sqrt{3}+\sqrt{6})/24$	0.24639964
$2T_2\ T_1T_2$	$(1+\sqrt{2})/8$	0.30177670	$-(1+\sqrt{2})/8$	-0.30177670
$2T_2\ 1T_2T_2$	$-\sqrt{2}/16$	-0.08838835	$-(4-\sqrt{2})/16$	-0.16161165
$2T_2\ 2T_2T_2$	$(4-\sqrt{2})/16$	0.16161165	$\sqrt{2}/16$	0.08838835

Table 20. Polyhedral isoscalars $\text{PIs}^2\begin{pmatrix}\Delta^A S^A A \\ \alpha\ \beta \\ a\ b\ c\end{pmatrix}$

$\alpha a\ \beta b\ c$	algebraic	numerical	$\alpha a\ \beta b\ c$	algebraic	numerical
$A_1\ A_1A_1$	$\sqrt{3}/12$	0.14433757	$1E\ 2T_2T_2$	$-(2-\sqrt{2})/16$	-0.03661165
$A_1\ 1T_2T_2$	$-\sqrt{6}/24$	-0.10206207	$1E\ T_1T_2$	$-\sqrt{3}/24$	-0.07216878
$A_1\ 2T_2T_2$	$\sqrt{6}/24$	0.10206207	$2E\ E\ A_1$	$\sqrt{6}/24$	0.10206207
$A_2\ T_1T_2$	$\sqrt{6}/12$	0.20412415	$2E\ 1T_2T_2$	$(\sqrt{2}-3)/8\sqrt{6}$	-0.08092433
$1E\ E\ A_1$	$\sqrt{2}/8$	0.17677670	$2E\ 2T_2T_2$	$-(\sqrt{2}+3)/8\sqrt{6}$	-0.22526189
$1E\ 1T_2T_2$	$(2+\sqrt{2})/16$	0.21338835	$2E\ T_1T_2$	$1/8$	0.12500000

Table 20. (continued)

αa βb c	algebraic	numerical
$1T_1$ T_1A_1	$(1+2\sqrt{3})/24$	0.18600423
$1T_1$ E T_2	$(-1+2\sqrt{3})/24$	0.10267090
$1T_1$ T_1T_2	$(2\sqrt{10}-\sqrt{2}-2\sqrt{6})/48$	0.00023671
$1T_1 1T_2T_2$	$(-1-2\sqrt{6}-2\sqrt{5}+\sqrt{2}-2\sqrt{3})/48$	-0.25877091
$1T_1 2T_2T_2$	$(-1+2\sqrt{6}-2\sqrt{5}-\sqrt{2}-2\sqrt{3})/48$	-0.11357233
$2T_1$ T_1A_1	$\sqrt{15}/24$	0.16137431
$2T_1$ E T_2	$-\sqrt{15}/24$	-0.16137431
$2T_1$ T_1T_2	$-(\sqrt{30}+2\sqrt{6})/48$	-0.21617094
$2T_1 1T_2T_2$	$(-\sqrt{15}+2\sqrt{3}+\sqrt{30})/48$	0.10559050
$2T_1 2T_2T_2$	$(-\sqrt{15}+2\sqrt{3}-\sqrt{30})/48$	-0.12262724
$3T_1$ T_1A_1	$(\sqrt{2}-\sqrt{6})/24$	-0.04313651
$3T_1$ E T_2	$-(\sqrt{2}+\sqrt{6})/24$	-0.16098764
$3T_1$ T_1T_2	$(-1+2\sqrt{5}+\sqrt{3})/24$	0.21684112
$3T_1 1T_2T_2$	$(2\sqrt{3}-\sqrt{2}-2\sqrt{10}+2+\sqrt{6})/48$	0.00364213
$3T_1 2T_2T_2$	$(2\sqrt{3}+\sqrt{2}+2\sqrt{10}+2-\sqrt{6})/48$	0.22402877
$1T_2 1T_2A_1$	$(1+\sqrt{3}+\sqrt{10})/24$	0.24559702
$1T_2 2T_2A_1$	$(1+\sqrt{3}-\sqrt{10})/24$	-0.01792612
$1T_2$ A_1T_2	$(\sqrt{3}-3-\sqrt{15})/36$	-0.14280368
$1T_2$ E T_2	$(\sqrt{6}-3\sqrt{2}+2\sqrt{30})/72$	0.12724028
$1T_2$ T_1T_2	$-(1+\sqrt{3})/24$	0.11383545
$1T_2 1T_2T_2$	$(\sqrt{2}+2-2\sqrt{3}+\sqrt{6})/48$	0.04999170
$1T_2 2T_2T_2$	$(\sqrt{2}-2+2\sqrt{3}+\sqrt{6})/48$	0.11099594
$2T_2 1T_2A_1$	$(\sqrt{30}-2\sqrt{3})/48$	0.04194008
$2T_2 2T_2A_1$	$(\sqrt{30}+2\sqrt{3})/48$	0.18627765
$2T_2$ A_1T_2	$(\sqrt{10}+\sqrt{2})/24$	0.19068713
$2T_2$ E T_2	$(\sqrt{5}-2)/24$	0.00983617
$2T_2$ T_1T_2	$\sqrt{30}/48$	0.11410887
$2T_2 1T_2T_2$	$(-\sqrt{15}-\sqrt{30})/48$	-0.19479602
$2T_2 2T_2T_2$	$(-\sqrt{15}+\sqrt{30})/48$	0.03342171
$3T_2 1T_2A_1$	$(\sqrt{2}-2\sqrt{6}+2\sqrt{5})/48$	0.02057021
$3T_2 2T_2A_1$	$(\sqrt{2}-2\sqrt{6}-2\sqrt{5})/48$	0.16576879
$3T_2$ A_1T_2	$(\sqrt{6}+6\sqrt{2}-\sqrt{30})/72$	0.07579924
$3T_2$ E T_2	$(\sqrt{3}+6+2\sqrt{15})/72$	0.21497247
$3T_2$ T_1T_2	$(-\sqrt{2}+2\sqrt{6})/48$	0.07259929
$3T_2 1T_2T_2$	$(1+\sqrt{2}+2\sqrt{6}-2\sqrt{3})/48$	0.08018940
$3T_2 2T_2T_2$	$(1-\sqrt{2}-2\sqrt{6}-2\sqrt{3})/48$	-0.18286031

23. Case study: Matrix elements

In this section, we apply the structural coefficients of the preceding section to the reduction of matrix elements. We begin with relations referring to no special type of AOs and procede step by step from the reduced matrix elements of the s.-a. MOs to the rotational invariants belonging to the integrals of GTOs. Since the representation of one AO by a sum of GTOs having the same angular momentum quantum number only enhances the complexity without further systematic insight, we shall confine each AO to one GTO. Because of clearness, we also take the simple example of the molecule P_4. The MOs of this molecule involve s- and p-orbitals:

$$\langle \vec{r} | Ai000 \rangle = \sqrt{1/4\pi} \cdot \exp(-\alpha^2 | \vec{r} - \vec{A_i} |^2) \tag{23.1}$$

$$\langle \vec{r} | Ai01m \rangle = -2i \cdot \exp(-\alpha^2 | \vec{r} - \vec{A_i} |^2) \cdot \alpha | \vec{r} - \vec{A_i} | \cdot Y_{1m}((\vec{r} - \vec{A_i})/| \vec{r} - \vec{A_i} |) \tag{23.2}$$

These orbitals transform according to the representations A_1 and T_2 of the group T_d:

$$| Ai00A_1i \rangle = | Ai000 \rangle \tag{23.3}$$

$$| Ai01T_2p \rangle = \sum_m | Ai01m \rangle \langle m | 1T_2p \rangle \tag{23.4}$$

From these AOs we build up the s.-a. MOs

$$| (Ae,00A_1)ep_e \rangle = \sum_i (\vec{A_i} | Aep_e) \cdot | Ai00A_1i \rangle \tag{23.5}$$

with $e=A_1$ or T_2 according to (22.29), and

$$| (Ae,01T_2)cp_c \rangle = \sum_{ip} K(cp_c, Aie, T_2p) \cdot | Ai01T_2p \rangle \tag{23.6}$$

with $e=A_1$ or T_2 again and c from the product $e \times T_2$. This means $c=T_2$ if $e=A_1$ and $c=A_1$, E, T_1, T_2 if $e=T_2$. The coefficients are given in table 8.

23.1. Step one: From the reduced matrix elements to BRMs

We now have to calculate three types of invariants of the molecular Hamiltonian $H=T+Q_A \cdot V_A$. The potential operator is defined in (8.10). The invariants are:

1) $\langle (Ae,00A_1)e \| H \| (Ae,00A_1)e \rangle$
2) $\langle (Ae,01T_2)c \| H \| (Af,01T_2)c \rangle$
3) $\langle (Ae,01T_2)c \| H \| (Ac,00A_1)c \rangle$

Since the Hamiltonian is a scalar operator, (5.13) applies reading now:

$$\langle (Ae,n_a l_a a)c \| H \| (Af,n_b l_b b)c \rangle \tag{23.7}$$
$$= \sum_{S\sigma k} GEO_1(Aa,Ab;ec,fc;S\sigma k)(An_a l_a a \| H \| An_b l_b b)_{S\sigma k}$$

with the geometrical factors

$$GEO_1(Aa,Ab;ec,fc;S\sigma k) \tag{23.8}$$
$$= \{c^+ fb\}\{ab^+ k\}\sqrt{Z(-AAS)dimc/dimk} \cdot \left\{ \begin{matrix} e & f & k \\ b^+ & a^+ & c^+ \end{matrix} \right\} \cdot PIs \begin{pmatrix} -A & A & S \\ & & \sigma \\ e^+ f & k \end{pmatrix} .$$

Of course, these factors are in dependent of the special AO quantum numbers n_a, l_a, n_b, l_b, and of the exponential parameters.

The involved BRMs are

in case 1): $(AOOA_1 \| H \| AOOA_1)_{Sk}$ with $Sk = 0^A A_1$ and $S^A A_1$,

in case 2): $(AO1T_2 \| H \| AO1T_2)_{Sok}$ with $Sok = 0^A A_1$, $0^A T_2$, $S^A A_1$, $S^A T_1$, $S^A 1T_2$, and $S^A 2T_2$,

in case 3): $(AO1T_2 \| H \| AOOA_1)_{Sok}$ with $Sok = 0^A T_2$, $S^A 1T_2$, and $S^A 2T_2$.

The geometrical factors of the relation (23.7), calculated by (23.8), are listed in the tables 21-23.

Table 21. The geometrical factors $GEO_1(AA_1, AA_1; ec, fc; Sk)$

e c , f c / Sk	$0^A A_1$	$S^A A_1$
$A_1 A_1$, $A_1 A_1$	1/2	$\sqrt{3}/2$
$T_2 T_2$, $T_2 T_2$	$\sqrt{3}/2$	-1/2

Table 22. The geometrical factors $GEO_1(AT_2 AT_2; ec, fc; Sok)$

e c , f c / Sok	$0^A A_1$	$0^A T_2$	$S^A A_1$	$S^A E$	$S^A T_1$	$S^A 1T_2$	$S^A 2T_2$
$T_2 A_1, T_2 A_1$	$\sqrt{3}/6$	$-\sqrt{2}/6$	$-1/6$	$-1/3$	$1/3$	$1/6$	$-1/6$
$T_2 E, T_2 E$	$\sqrt{6}/6$	$1/6$	$-\sqrt{2}/6$	$-\sqrt{2}/3$	$-\sqrt{2}/6$	$-\sqrt{2}/12$	$\sqrt{2}/12$
$T_2 T_1, T_2 T_1$	$1/2$	$\sqrt{6}/12$	$-\sqrt{3}/6$	$\sqrt{3}/6$	$\sqrt{3}/6$	$-\sqrt{3}/12$	$\sqrt{3}/12$
$T_2 T_2, T_2 T_2$	$1/2$	$-\sqrt{6}/12$	$-\sqrt{3}/6$	$\sqrt{3}/6$	$-\sqrt{3}/6$	$\sqrt{3}/12$	$-\sqrt{3}/12$
$A_1 T_2, A_1 T_2$	$1/2$	0	$\sqrt{3}/2$	0	0	0	0
$A_1 T_2, T_2 T_2$	0	$\sqrt{3}/6$	0	0	0	$\frac{\sqrt{2}-2}{4\cdot\sqrt{3}}$	$-\frac{\sqrt{2}+2}{4\cdot\sqrt{3}}$
$T_2 T_2, A_1 T_2$	0	$\sqrt{3}/6$	0	0	0	$\frac{\sqrt{2}+2}{4\cdot\sqrt{3}}$	$-\frac{\sqrt{2}-2}{4\cdot\sqrt{3}}$

Table 23. The geometrical factors $GEO_1(AT_2, AA_1; ec, fc; Sok)$

e c , f c / Sok	$0^A T_2$	$S^A 1T_2$	$S^A 2T_2$
$T_2 A_1, A_1 A_1$	$\sqrt{3}/6$	$(\sqrt{2}+2)/4\sqrt{3}$	$-(\sqrt{2}-2)/4\sqrt{3}$
$A_1 T_2, T_2 T_2$	$\sqrt{3}/6$	$(\sqrt{2}-2)/4\sqrt{3}$	$-(\sqrt{2}+2)/4\sqrt{3}$
$T_2 T_2, T_2 T_2$	$-\sqrt{6}/6$	$\sqrt{3}/6$	$-\sqrt{3}/6$

Using these factors, an example of the relation (23.7) is given by:

$$\langle (AT_2, 01T_2)T_1 \| H \| (AT_2, 01T_2)T_1 \rangle = (1/2)\cdot(AO1T_2 \| H \| AO1T_2)_0{}^A A_1$$
$$+(\sqrt{6}/12)\cdot(AO1T_2 \| H \| AO1T_2)_0{}^A T_2 -(\sqrt{3}/6)\cdot(AO1T_2 \| H \| AO1T_2)_S{}^A A_1$$
$$+(\sqrt{3}/6)\cdot(AO1T_2 \| H \| AO1T_2)_S{}^A E +(\sqrt{3}/6)\cdot(AO1T_2 \| H \| AO1T_2)_S{}^A T_1 \qquad (23.9)$$
$$-(\sqrt{3}/12)\cdot(AO1T_2 \| H \| AO1T_2)_S{}^A 1T_2 +(\sqrt{3}/12)\cdot(AO1T_2 \| H \| AO1T_2)_S{}^A 2T_2$$

The same analysis applies to the reduced matrix elements of the overlap matrix. We only have to cancel the Hamiltonian in all relations of this subsection. The energy eigenvalues of the symmetry species E and T_1 then simply are:

$$E(c) = \frac{\langle (AT_2,01T_2)c\|H\|(AT_2,01T_2)c\rangle}{\langle (AT_2,01T_2)c\|(AT_2,01T_2)c\rangle} \quad \text{with } c=E, T_1 \qquad (23.10)$$

Since the species A_1 and T_2 occur twice and thrice, there remains a two- or a three-dimensional eigenvalue problem in these cases.

23.2. Step two: From BRMs to TRMs

The improper BRMs of the potential operator V_A are further related to the TRMs of the triangles Δ_0^A, Σ_0^A, Σ_1^A, Σ_2^A, and Δ^A by formula (8.14). More precisely we have the following relations (23.11) and (23.12):

$$(An_a1_aa\|V_A\|An_b1_bb)_0A_k$$
$$= \sum_{\Delta\gamma} \sqrt{4\cdot Z(\Delta)}\cdot PIs^2 \begin{pmatrix} \Delta & 0^A A \\ \gamma & \\ k & k & A_1 \end{pmatrix} \cdot (An_a1_aa\|Ar^{-1}\|An_b1_bb)_{\Delta\gamma k} \, , \qquad (23.11)$$

where the summands are determined by the (degenerate) triangles sharing the (degenerate) edge 0^A, i.e. $\Delta\gamma = \Delta_0^A$, Σ_0^A1, and Σ_0^A2. The coefficients of this relation are listed in table 24.

Table 24. The factors $\sqrt{4\cdot Z(\Delta)}\cdot PIs^2 \begin{pmatrix} \Delta & 0^A A \\ \gamma & \\ k & k & A_1 \end{pmatrix}$

$\Delta\gamma$ k	Δ_0^A	Σ_0^A1	Σ_0^A2
A_1	1	$\sqrt{3}$	-
T_2	$\sqrt{3}$	$(\sqrt{3}+\sqrt{6})/\sqrt{2}$	$(\sqrt{3}-\sqrt{6})/\sqrt{2}$

The corresponding relation referring to the edge s^A is given by:

$$(An_a1_aa\|V_A\|An_b1_bb)_{s^A0k}$$
$$= \sum_{\Delta\gamma} \sqrt{12\cdot Z(\Delta)}\cdot PIs^2 \begin{pmatrix} \Delta & s^A A \\ \gamma & \sigma \\ k & k & A_1 \end{pmatrix} \cdot (An_a1_aa\|Ar^{-1}\|An_b1_bb)_{\Delta\gamma k} \, , \qquad (23.12)$$

where now $\Delta = \Sigma_1^A$, Σ_2^A, Δ^A and $\gamma=1, 2, 3$. The coefficients of this relation are listed in table 25.

Table 25. The factors $\sqrt{12\cdot Z(\Delta)}\cdot PIs^2 \begin{pmatrix} \Delta & s^A A \\ \gamma & \sigma \\ k & k & A_1 \end{pmatrix}$

$\Delta\gamma$ $\sigma\gamma k$	Σ_1^A	Σ_2^A	Δ^A
A_1	1	1	$\sqrt{2}$
1E	$\sqrt{2}$	$\sqrt{2}$	$\sqrt{3}$
2E	-	-	1
$1T_1$	$\sqrt{3}$	$-\sqrt{3}$	$(1+\sqrt{12})/\sqrt{6}$
$2T_1$	-	-	$2\sqrt{15}$

Table 25. (continued)

σγk \ Δγ	Σ_1^A	Σ_2^A	Δ^A
$3T_1$	-	-	$(\sqrt{3}-3)/3$
$11T_2$	$\sqrt{3}$	0	$(1+\sqrt{3}+\sqrt{10})/\sqrt{6}$
$12T_2$	0	$-\sqrt{3}$	$(\sqrt{5}-\sqrt{2})/\sqrt{6}$
$13T_2$	-	-	$(1-\sqrt{12}+\sqrt{10})/\sqrt{12}$
$21T_2$	0	$-\sqrt{3}$	$(1+\sqrt{3}-\sqrt{10})/\sqrt{6}$
$22T_2$	$\sqrt{3}$	0	$(\sqrt{5}+\sqrt{2})/\sqrt{6}$
$23T_2$	-	-	$(1-\sqrt{12}-\sqrt{10})/\sqrt{12}$

Applying these results, we give an example of (23.12):

$$(A01T_2\|V_A\|A01T_2)_S{}^A{}_{T_1} = \sqrt{3}\cdot(A01T_2\|Ar^{-1}\|A01T_2)_{\Sigma_1}{}^A{}_{T_1} \qquad (23.13)$$

$$-\sqrt{3}\cdot(A01T_2\|Ar^{-1}\|A01T_2)_{\Sigma_2}{}^A{}_{T_1} +((1+\sqrt{12})/\sqrt{6})\cdot(A01T_2\|Ar^{-1}\|A01T_2)_\Delta{}^A{}_{1T_1}$$

$$+2\sqrt{15}\cdot(A01T_2\|Ar^{-1}\|A01T_2)_\Delta{}^A{}_{2T_1} +((1-\sqrt{3})/\sqrt{3})\cdot(A01T_2\|Ar^{-1}\|A01T_2)_\Delta{}^A{}_{3T_1}$$

23.3. Step three: Ab initio calculation of BRMs

In the case of proper BRMs, we can relate them to the rotational invariants of the two-centre integrals according to formula (15.2). Because of (12.14), this formula reads for a scalar operator:

$$(An_al_aa\|T\|An_bl_bb)_{S\varepsilon e} = (-1)^{l_a+l_b}\cdot\sqrt{4\pi\cdot\dime} \qquad (23.14)$$

$$\cdot\sum_j c(S\varepsilon e,j)\cdot S^j\cdot Is\begin{pmatrix}1 & 1 & j\\ a & b & \\ a & b & e\end{pmatrix}\cdot\langle n_al_a\|-S^j,T\|n_bl_b\rangle^j$$

The expansion coefficients $c(S\varepsilon e,j)$ are listed in table 26.

Table 26. The coefficients $c(S\varepsilon e,j)$

$S\ \varepsilon e\ ,j$	$c(S\varepsilon e,j)$	$S\ \varepsilon e\ ,j$	$c(S\varepsilon e,j)$
$0^A\ A_1,0$	$2/\sqrt{4\pi}$	$S^A2T_2,1$	$i\sqrt{6}/4\pi$
$0^A\ e\ ,j>0$	0	$S^A\ E\ ,2$	$\sqrt{15}/8\pi$
$S^A\ A_1,0$	$\sqrt{12}/4\pi$	$S^A1T_2,2$	$-\sqrt{15}/8\pi$
$S^A1T_2,1$	$i\sqrt{6}/4\pi$	$S^A2T_2,2$	$\sqrt{15}/8\pi$

The special cases of (23.14) are:

$$\left.\begin{array}{l}(A00A_1\|T\|A00A_1)_0{}^A{}_{A_1} = 2\cdot\langle 00\|0^\circ,T\|00\rangle^\circ\\[4pt](A00A_1\|T\|A00A_1)_S{}^A{}_{A_1} = \sqrt{12}\cdot\langle 00\|-S^{Ao},T\|00\rangle^\circ\\[4pt](A01T_2\|T\|A00A_1)_0{}^A{}_{T_2} = 0\end{array}\right\} \quad (23.15)$$

$$\left.\begin{array}{l}(A01T_2\|T\|A00A_1)_S{}^A{}_{1T_2} = -3\sqrt{2}\cdot\langle 01\|-S^{A1},T\|00\rangle^1\cdot S^A\\[4pt](A01T_2\|T\|A00A_1)_S{}^A{}_{2T_2} = 3\sqrt{3}\cdot\langle 01\|-S^{A1},T\|00\rangle^1\cdot S^A\end{array}\right\} \quad (23.16)$$

$$
\left.
\begin{aligned}
(A01T_2\|T\|A01T_2)_0{}^A{}_{A_1} &= 2\cdot\langle 01\|0^{0},T\|01\rangle^{0}\\
(A01T_2\|T\ A01T_2)_0{}^A{}_{T_2} &= 0\\
(A01T_2\|T\|A01T_2)_S{}^A{}_{A_1} &= \sqrt{12}\cdot\langle 01\|-s^{A0},T\|01\rangle^{0}\\
(A01T_2\|T\|A01T_2)_S{}^A{}_{E} &= \sqrt{6}\cdot\langle 01\|-s^{A2},T\|01\rangle^{2}\cdot(s^{A})^{2}\\
(A01T_2\|T\|A01T_2)_S{}^A{}_{T_1} &= 0\\
(A01T_2\|T\|A01T_2)_S{}^A{}_{1T_2} &= -3\sqrt{6}\langle 01\|-s^{A2},T\|01\rangle^{2}\cdot(s^{A})^{2}\\
(A01T_2\|T\|A01T_2)_S{}^A{}_{2T_2} &= 3\sqrt{6}\langle 01\|-s^{A2},T\|01\rangle^{2}\cdot(s^{A})^{2}
\end{aligned}
\right\}
\tag{23.17}
$$

Again, the same relations apply to the overlap matrix, if we omit the operator T. In both cases, the matrix elements are reduced to six independent parameters, the rotational invariants.

Up to this point, we have made no reference to a special system of orbitals. In order to calculate the rotational invariants, we do this now and choose the Gauss type orbitals of ref.[46]. By comparing formula (13.3) to formula (4.1) of [46], we conclude:

$$
\langle n_a l_a\|AB^{j}\|n_b l_b\rangle^{j} = \sqrt{1/4\pi}\cdot(-1)^{l_a+l_b+j}\xi^{j}_{\alpha\beta}\cdot\varepsilon^{0}_{\alpha\beta}(n_a n_b, j l_a l_b, AB)\ ,
\tag{23.18}
$$

where α and β are the orbital exponents of the bra- and ket-orbitals respectively. We use the following abbreviations:

$$
\xi_{\alpha\beta} = \alpha\beta/\sqrt{\alpha^2+\beta^2},\qquad \theta_{\alpha\beta} = \sqrt{\alpha^2+\beta^2},\qquad \varphi_{\alpha\beta} = \arctan(\alpha/\beta)
\tag{23.19}
$$

In the same way, the comparison of (13.3) to formula (4.3) of [46] yields the rotational invariants of the kinetic energy:

$$
\langle n_a l_a\| AB^{j},T\|n_b l_b\rangle^{j}
$$
$$
= -(\hbar^2\beta^2/2m)\sqrt{1/4\pi}\cdot(-1)^{l_a+l_b+j}\xi^{j}_{\alpha\beta}\cdot\varepsilon^{0}_{\alpha\beta}(n_a n_b+1, j l_a l_b, AB)
\tag{23.20}
$$

According to (4.2) of [46], the functions $\varepsilon^{0}_{\alpha\beta}(...)$ are given by:

$$
\varepsilon^{0}_{\alpha\beta}(n_a n_b, j l_a l_b, AB)
\tag{23.21}
$$
$$
= [Nj,00,j\|\varphi_{\alpha\beta}\|n_a l_a, n_b l_b, j]\cdot(\pi/2\theta^{3}_{\alpha\beta}\sqrt{2j+1})\cdot\varphi^{0}(\xi_{\alpha\beta},Nj,AB)
$$

with $N = n_a+n_b+(l_a+l_b-j)/2$. Using the coefficients and Gauss-Laguerre type functions listed in appendix 3, we obtain the following six invariants:

$$
\begin{aligned}
\langle 00\|0^{0}\|00\rangle^{0} &= \sqrt{\pi}/4\theta^{3}_{\alpha\beta}\\
\langle 00\|-s^{A0}\|00\rangle^{0} &= (\sqrt{\pi}/4\theta^{3}_{\alpha\beta})\cdot\exp(-\xi^{2}_{\alpha\beta}s^{A2})\\
\langle 01\|-s^{A1}\|00\rangle^{1} &= -(\sqrt{3\pi}\beta\,\xi_{\alpha\beta}/2\theta^{4}_{\alpha\beta})\cdot\exp(-\xi^{2}_{\alpha\beta}s^{A2})\\
\langle 01\|0^{0}\|01\rangle^{0} &= 3\sqrt{3\pi}\,\xi_{\alpha\beta}/2\theta^{4}_{\alpha\beta}\\
\langle 01\|-s^{A0}\|01\rangle^{0} &= (\sqrt{3\pi}\,\xi_{\alpha\beta}/2\theta^{4}_{\alpha\beta})\cdot(3-2\xi^{2}_{\alpha\beta}s^{A2})\cdot\exp(-\xi^{2}_{\alpha\beta}s^{A2})\\
\langle 01\|-s^{A2}\|01\rangle^{2} &= -(\sqrt{6\pi}\xi^{3}_{\alpha\beta}/\theta^{4}_{\alpha\beta})\cdot\exp(-\xi^{2}_{\alpha\beta}s^{A2})
\end{aligned}
\tag{23.22}
$$

And with respect to the kinetic energy, we get:

$$\langle 00\|0^{o},T\|00\rangle^{o} \;=\; 3\sqrt{\pi}\hbar^{2}\,\xi_{\alpha\beta}^{2}/4m\theta_{\alpha\beta}^{3}$$

$$\langle 00\| -s^{Ao},T\|00\rangle^{o} \;=\; (\sqrt{\pi}\hbar^{2}\,\xi_{\alpha\beta}^{2}/4m\theta_{\alpha\beta}^{3})\cdot(3-2\xi_{\alpha\beta}^{2}s^{A2})\cdot\exp(-\xi_{\alpha\beta}^{2}s^{A2})$$

$$\langle 01\| -s^{A1},T\|00\rangle^{1} \;=\; -(\sqrt{\pi}\hbar^{2}\,\xi_{\alpha\beta}^{3}\beta/2m\theta_{\alpha\beta}^{4})\cdot(5-2\xi_{\alpha\beta}^{2}s^{A2})\cdot\exp(-\xi_{\alpha\beta}^{2}s^{A2})$$

$$\langle 01\|0^{o},T\|01\rangle^{o} \;=\; 15\sqrt{3}\hbar^{2}\,\xi_{\alpha\beta}^{3}/2m\theta_{\alpha\beta}^{4} \tag{23.23}$$

$$\langle 01\| -s^{Ao},T\|01\rangle^{o} \;=\; (\sqrt{3}\hbar^{2}\,\xi_{\alpha\beta}^{3}/2m\theta_{\alpha\beta}^{4})\cdot(15-20\xi_{\alpha\beta}^{2}s^{A2}+4\xi_{\alpha\beta}^{4}s^{A4})\cdot\exp(-\xi_{\alpha\beta}^{2}s^{A2})$$

$$\langle 01\| -s^{A2},T\|01\rangle^{2} \;=\; -(\sqrt{6}\hbar^{2}\,\xi_{\alpha\beta}^{5}/m\theta_{\alpha\beta}^{4})\cdot(7-2\xi_{\alpha\beta}^{2}s^{A2})\cdot\exp(-\xi_{\alpha\beta}^{2}s^{A2})$$

We finally come back to the example (23.9). Inserting (23.17 and 23) into (23.9) yields the reduced matrix element of the kinetic energy with respect·to the s.-a. MO of species T_{1}:

$$\langle\!\langle AT_{2},01T_{2})T_{1}\| T \|(AT_{2},01T_{2})T_{1}\rangle\!\rangle$$

$$= \langle 01\|0^{o},T\|01\rangle^{o} - \langle 01\| -s^{Ao},T\|01\rangle^{o} + \sqrt{1/2}\,\langle 01\| -s^{A2},T\|01\rangle^{2}\cdot s^{A2}$$

$$= (\sqrt{3}\hbar^{2}\,\xi_{\alpha\beta}^{2}/2m\theta_{\alpha\beta}^{4})\cdot\Big[15-(15-6\xi_{\alpha\beta}^{2}s^{A2})\cdot\exp(-\xi_{\alpha\beta}^{2}s^{A2})\Big] \tag{23.24}$$

The corresponding element of the overlap matrix is:

$$\langle\!\langle AT_{2},01T_{2})T_{1}\|(AT_{2},01T_{2})T_{1}\rangle\!\rangle = (3\sqrt{3}\pi\,\xi_{\alpha\beta}^{2}/2\theta_{\alpha\beta}^{4})\cdot\big[1-\exp(-\xi_{\alpha\beta}^{2}s^{A2})\big] \tag{23.25}$$

23.4. Step four: The ab initio calculation of the TRMs

There finally remains the calculation of the TRMs of the nuclear attraction. Working in the AB–PC scheme, we have the following relation of TRMs and rotational invariants, which is quite analogous to (15.14/15):

$$(A\varphi_{a}a\| Cr^{-1}\| B\varphi_{b}b)_{\Delta\gamma c}^{\varepsilon}$$

$$= \sum_{JjL}GEO_{5}'(Al_{a}\alpha a,Bl_{b}\beta b;C\Delta\gamma c;JjL)\,\langle An_{a}l_{a}\|AB_{\Delta}^{J}PC_{\Delta}^{j}\|Bn_{b}l_{b}\rangle^{L}, \tag{23.26}$$

where the geometrical factor is given by:

$$GEO_{5}'(Al_{a}\alpha a,Bl_{b}\beta b;C\Delta\gamma c;JjL)= 4\pi(2L+1)\sqrt{1/dimc}\sum_{\gamma\delta d\eta e\mu\sigma\tau}\bar{c}^{2}(\Delta\gamma c,st.\delta d\eta e\gamma')$$

$$\cdot Is\begin{pmatrix}1^{+}1_{b}L^{+}\\ \alpha^{a}\beta^{b}\mu\\ a^{+}b\ c^{+}\end{pmatrix}_{\varepsilon} Is\begin{pmatrix}L\ J^{+}j^{+}\\ \mu\ \sigma\ \tau\\ c\ d^{+}e^{+}\end{pmatrix}_{\gamma'}\cdot S_{\delta d}^{J\sigma}(AB_{\Delta})S_{\eta e}^{j\tau}(PC_{\Delta}) \tag{23.27}$$

With respect to the group T_{d} and to the low angular momentum quantum numbers, several multiplicity indices become obsolete. Further, all centres involved are of type A:

$$(An_{a}l_{a}a\| Ar^{-1}\| An_{b}l_{b}b)_{\Delta\gamma c}= \sum_{JjL}GEO_{5}'(Al_{a}a,Al_{b}b;A\Delta\gamma c;JjL)$$

$$\cdot\langle An_{a}l_{a}\| AA_{\Delta}^{J}PA_{\Delta}^{j}\|An_{b}l_{b}\rangle^{L} \tag{23.28}$$

with

$$GEO_{5}'(Al_{a}a,Al_{b}b;A\Delta\gamma c;JjL) \tag{23.29}$$

$$= 4\pi(2L+1)\sqrt{1/dimc}\sum_{\delta d\eta e}Is(\begin{smallmatrix}1^{+}1_{b}L^{+}\\ a\ b\ c^{+}\end{smallmatrix})Is(\begin{smallmatrix}LJ^{+}j^{+}\\ cd^{+}e^{+}\end{smallmatrix})\bar{c}^{2}(\Delta\gamma c,st.\delta d\eta e)S_{\delta d}^{J}(AA_{\Delta})S_{\eta e}^{j}(PA_{\Delta})$$

In order to save space, it is advisable to compile the compound coefficients

$$c^2(\Delta\gamma c,(Jj)Lc)$$
$$= \sqrt{(2L+1)/dimc} \cdot \sum_{\delta d\eta e} Is(\overset{LJ^+j^+}{cd}\overset{}{+}\overset{}{e}^+) c^2(\Delta\gamma c,st.\delta d\eta e) S_{\delta d}^J(AA_\Delta) S_{\eta e}^j(PA_\Delta). \tag{23.30}$$

These coefficients can be calculated directly:

$$c^2(\Delta\gamma c,(Jj)Lc) = \sqrt{(2L+1)Z(\Delta)} dimc^{-1} \sum_{de} Is(\overset{LJ^+j^+}{cd}\overset{}{+}\overset{}{e}^+) \sum_{jrm} \tau(\overset{\Delta SA}{jrm})(\overset{c\ d^+e^+}{P_cP_dP_e}) \tag{23.31}$$
$$\cdot (\Delta\gamma cp_c|\Delta_j)\langle\overrightarrow{S_r}|sol\ Jdp_d\rangle\langle\overrightarrow{P_rA_m}|sol\ jep_e\rangle$$

According to section 13.3., the vector $\overrightarrow{P_r}$ belonging to $\overrightarrow{S_r}$ is determined by the topological correlation

$$\overrightarrow{P_r} = \sqrt{Z(S)} \sum_{kl} \tau(\overset{-AAS}{klr}) \cdot (\alpha^2\overrightarrow{A_k}+\beta^2\overrightarrow{A_1})/(\alpha^2+\beta^2) \tag{23.32}$$

The isoscalar factors and s.-a. solid harmonics for this calculation are listed in appendix 4. The results can be checked by the sum rule:

$$\sum_{L\gamma c} dimc|c^2(\Delta\gamma c,(Jj)Lc)|^2 = Z(\Delta)(AA_\Delta)^{2J}(PA_\Delta)^{2j}(2J+1)(2j+1)/16\pi^2 \tag{23.33}$$

We arrange the results according to the triangles involved. Since the branching rules of the group chain $R(3) \supset T_d$ depend on the inversion property of the representations, we add the indices g or u to the quantum number L.

1) The triangle is Δ_o^A. This means: $\Delta=\Delta_o^A$, $S=0^A$, $\Lambda A_\Delta=0$, $PA_\Delta=0$.

$$c^2(\Delta_o^A A_1,(00)0_g A_1) = 1/2\pi \tag{23.34}$$

2) The triangle is Σ_o^A. This means: $\Delta=\Sigma_o^A$, $S=0^A$, $AA_\Delta=0$, $PA_\Delta=S^A$.

$$c^2(\Sigma_o^A A_1,(00)0_g A_1) = \sqrt{3}/2\pi \qquad c^2(\Sigma_o^A 1T_2,(02)2_g T_2) = -\sqrt{30}S^{A2}/8\pi$$
$$c^2(\Sigma_o^A 1T_2,(01)1_u T_2) = -i\sqrt{6}S^A/4\pi \qquad c^2(\Sigma_o^A 2T_2,(02)2_g T_2) = \sqrt{30}S^{A2}/8\pi$$
$$c^2(\Sigma_o^A 2T_2,(01)1_u T_2) = -i\sqrt{6}S^A/4\pi \qquad c^2(\Sigma_o^A E,(02)2_g E) = \sqrt{30}S^{A2}/8\pi \tag{23.35}$$

3) The triangle is Σ_1^A. This means: $\Delta=\Sigma_1^A$, $S=S^A$, $AA_\Delta=S^A$, $PA_\Delta=\beta^2 S^A/(\alpha^2+\beta^2)$.

Table 27. The coefficients $c^2(\Sigma_1^A\gamma c,(Jj)Lc)$.

J	j	L	γc	c²	J	j	L	γc	c²
0	0	0$_g$	A$_1$	$\sqrt{3}/2\pi$	1	1	2$_g$	2T$_2$	$-3\beta^2 S^{A2}/4\pi(\alpha^2+\beta^2)$
1	0	1$_u$	1T$_2$	$i\sqrt{6}S^A/4\pi$	1	1	2$_g$	E	$-3\beta^2 S^{A2}/4\pi(\alpha^2+\beta^2)$
1	0	1$_u$	2T$_2$	$i\sqrt{6}S^A/4\pi$	2	0	2$_g$	E	$\sqrt{30}S^{A2}/8\pi$
0	1	1$_u$	1T$_2$	$i\sqrt{6}\beta^2 S^A/4\pi(\alpha^2+\beta^2)$	2	0	2$_g$	1T$_2$	$-\sqrt{30}S^{A2}/8\pi$
0	1	1$_u$	2T$_2$	$i\sqrt{6}\beta^2 S^A/4\pi(\alpha^2+\beta^2)$	2	0	2$_g$	2T$_2$	$\sqrt{30}S^{A2}/8\pi$
1	1	0$_g$	A$_1$	$-3\beta^2 S^{A2}/2\pi(\alpha^2+\beta^2)$	0	2	2$_g$	E	$\sqrt{30}\beta^4 S^{A2}/8\pi(\alpha^2+\beta^2)^2$
1	1	1$_g$	T$_1$	0	0	2	2$_g$	1T$_2$	$-\sqrt{30}\beta^4 S^{A2}/8\pi(\alpha^2+\beta^2)^2$
1	1	2$_g$	1T$_2$	$3\beta^2 S^{A2}/4\pi(\alpha^2+\beta^2)$	0	2	2$_g$	2T$_2$	$\sqrt{30}\beta^4 S^{A2}/8\pi(\alpha^2+\beta^2)^2$

4) The triangle is Σ_2^A. This means: $\Delta=\Sigma_2^A$, $S=S^A$, $AA_\Delta=S^A$, $PA_\Delta=\alpha^2 S^A/(\alpha^2+\beta^2)$.

$$c^2(\Sigma_2^A\gamma c,(Jj)Lc) = (-\alpha^2/\beta^2)^j \cdot c^2(\Sigma_1^A\gamma c,(Jj)Lc) \tag{23.36}$$

5) The triangle is Δ^A. This means: $\Delta = \Delta^A$, $S = S^A$, $AA_\Delta = S^A$, and $PA_\Delta = S^A \cdot \sqrt{\alpha^4 + \beta^4 + \alpha^2\beta^2}/(\alpha^2 + \beta^2)$. The distances S^A (between two vertices) and A (between vertex and centre) are related by $A = S^A \cdot \sqrt{3/8}$.

Table 28. The coefficients $c^2(\Delta\gamma c,(Jj)Lc)$.

J j L γc	$c^2(\Delta\gamma c,(Jj)Lc)$
$0\ 0\ 0_g\ A_1$	$\sqrt{6}/4\pi$
$1\ 0\ 1_u\ 1T_2$	$i(\sqrt{3}+3)S^A/6\pi$
$1\ 0\ 1_u\ 2T_2$	$i\sqrt{10}S^A/4\pi$
$1\ 0\ 1_u\ 3T_2$	$i(\sqrt{6}-6\sqrt{2})S^A/12\pi$
$0\ 1\ 1_u\ 1T_2$	$iS^A(-\sqrt{15}(\alpha^2+\beta^2)+\sqrt{3}\beta^2-6\alpha^2)/6\pi(\alpha^2+\beta^2)$
$0\ 1\ 1_u\ 2T_2$	$iS^A(\sqrt{2}(\alpha^2+\beta^2)+\sqrt{10}\beta^2)/4\pi(\alpha^2+\beta^2)$
$0\ 1\ 1_u\ 3T_2$	$iS^A(-\sqrt{30}(\alpha^2+\beta^2)+\sqrt{6}\beta^2+6\sqrt{2}\alpha^2)/12\pi(\alpha^2+\beta^2)$
$1\ 1\ 0_g\ A_1$	$3\sqrt{2}S^{A2}(\alpha^2-\beta^2)/4\pi(\alpha^2+\beta^2)$
$1\ 1\ 1_g\ 1T_1$	$S^{A2}(\sqrt{10}+\sqrt{2}+2\sqrt{6})/8\pi$
$1\ 1\ 1_g\ 2T_1$	$S^{A2}(\sqrt{30}-\sqrt{6})/8\pi$
$1\ 1\ 1_g\ 3T_1$	$S^{A2}(\sqrt{5}+1-\sqrt{3})/4\pi$
$1\ 1\ 2_g\ 1E$	$3\sqrt{6}\beta^2 S^{A2}/8\pi(\alpha^2+\beta^2)$
$1\ 1\ 2_g\ 2E$	$3\sqrt{2}S^{A2}(2\alpha^2+\beta^2)/8\pi(\alpha^2+\beta^2)$
$1\ 1\ 2_g\ 1T_2$	$-S^{A2}(\sqrt{5}(\alpha^2-\beta^2)+(1+\sqrt{3})(\alpha^2+\beta^2))/4\pi(\alpha^2+\beta^2)$
$1\ 1\ 2_g\ 2T_2$	$S^{A2}(\sqrt{6}(\alpha^2-\beta^2)-\sqrt{30}(\alpha^2+\beta^2))/8\pi(\alpha^2+\beta^2)$
$1\ 1\ 2_g\ 3T_2$	$S^{A2}(-\sqrt{10}(\alpha^2-\beta^2)+(2\sqrt{6}-\sqrt{2})(\alpha^2+\beta^2))/8\pi(\alpha^2+\beta^2)$
$2\ 0\ 2_g\ 1E$	$\sqrt{15}S^{A2}/8\pi$
$2\ 0\ 2_g\ 2E$	$3\sqrt{5}S^{A2}/8\pi$
$2\ 0\ 2_g\ 1T_2$	$-5\sqrt{6}S^{A2}/12\pi$
$2\ 0\ 2_g\ 2T_2$	$\sqrt{5}S^{A2}/4\pi$
$2\ 0\ 2_g\ 3T_2$	$-5\sqrt{3}S^{A2}/12\pi$
$0\ 2\ 2_g\ 1E$	$-\sqrt{30}S^{A2}(\alpha^4+2\alpha^2\beta^2)/4\pi(\alpha^2+\beta^2)^2$
$0\ 2\ 2_g\ 2E$	$\sqrt{10}S^{A2}(\alpha^4-2\alpha^2\beta^2-2\beta^4)/4\pi(\alpha^2+\beta^2)^2$
$0\ 2\ 2_g\ 1T_2$	$S^{A2}(5\alpha^2\beta^2-\sqrt{5}\alpha^2(\alpha^2+\beta^2)+\sqrt{15}\beta^2(\alpha^2+\beta^2))/3\pi(\alpha^2+\beta^2)^2$
$0\ 2\ 2_g\ 2T_2$	$S^{A2}(\sqrt{30}\alpha^2\beta^2+5\sqrt{6}\alpha^2(\alpha^2+\beta^2))/6\pi(\alpha^2+\beta^2)^2$
$0\ 2\ 2_g\ 3T_2$	$S^{A2}(5\sqrt{2}\alpha^2\beta^2-3\sqrt{10}\alpha^2(\alpha^2+\beta^2)-6\sqrt{30}\beta^2(\alpha^2+\beta^2))/6\pi(\alpha^2+\beta^2)^2$

These last coefficients are rather complex. In order to avoid this complexity, one has to transform the three-centre integrals into a scheme of distance vectors, which is independent of the orbital exponents α and β.

With these results, we determine the geometrical factors. From (23. 29 and 30) follows:

$$GEO_5(Al_a a, Al_b b; A\Delta\gamma c; JjL) = 4\pi \cdot \sqrt{2L+1} \cdot Is(^{l_a^+ l_b^+ L^+}_{a\ b\ c}) \cdot c^2(\Delta\gamma c,(Jj)Lc) \qquad (23.37)$$

Since the isoscalar factors with $l_a, l_b \leq 1$ are very simple, we get the following equation:

$$GEO_5(Al_a a, Al_b b; A\Delta\gamma c; JjL) = 4\pi\sqrt{dimc} \cdot c^2(\Delta\gamma c,(Jj)Lc) \qquad (23.38)$$

We now come back to the example (23.13). Using the tables 27 and 28, we obtain:

$$(\text{AO1T}_2 \| \text{Ar}^{-1} \text{AAO1T}_2)_{\Sigma_1^A T_1} = 0, \quad (\text{AO1T}_2 \| \text{Ar}^{-1} \| \text{AO1T}_2)_{\Sigma_2^A T_1} = 0, \quad (23.39)$$

$$(\text{AO1T}_2 \| \text{Ar}^{-1} \| \text{AO1T}_2)_{\Delta^A \gamma T_1} = 4\pi\sqrt{3} \cdot c^2 (\Delta \gamma T_1, (11)1_g T_1) \langle \text{AO1} \| \text{AA}_\Delta^{\frac{1}{2}} \text{PA}_\Delta^{\frac{1}{2}} \| \text{AO1} \rangle^1 \tag{23.40}$$

Therefore the BRM of the operator V_A is given by one rotational invariant only:

$$(\text{AO1T}_2 \| V_A \| \text{AO1T}_2)_S A_{T_1} = (3S^{A2}/2)(\sqrt{5} + 7 + 10\sqrt{6} - 2\sqrt{30}) \langle \text{AO1} \| \text{AA}_\Delta^{\frac{1}{2}} \text{PA}_\Delta^{\frac{1}{2}} \| \text{AO1} \rangle^1 \tag{23.41}$$

As in the preceding subsection, we have not yet made reference to a special system of orbitals. We do this now in order to calculate the rotational invariants of eq.(23.28). For the Gauss-type orbitals of ref. [46] we have the general relation (A2.14). This relation now reads:

$$\langle \text{AO1}_a \| \text{AA}_\Delta^J \text{PA}_\Delta^j \| \text{AO1}_b \rangle^L = (4\pi/(2L+1)) \xi_{\alpha\beta}^J \theta_{\alpha\beta}^j \cdot \eta_{\alpha\beta}^0 (00, \text{Ll}_a 1_b, \text{Jj}, \text{AA}_\Delta, \text{PA}_\Delta) \quad (23.42)$$

According to (5.8) of [46] the functions $\eta_{\alpha\beta}^0$ are given by:

$$\eta_{\alpha\beta}^0 (00, \text{Ll}_a 1_a, \text{Jj}, \text{AA}_\Delta, \text{PA}_\Delta) = -4\pi\theta_{\alpha\beta}^{-2} \sum_{Nn} [\text{NJ}, nj, L \| \varphi_{\alpha\beta} \| 01_a, 01_b, L]$$

$$\cdot \varphi^0(\xi_{\alpha\beta}, \text{NJ}, \text{AA}_\Delta) \cdot \varphi^0(\theta_{\alpha\beta}, n-1j, \text{PA}_\Delta) \tag{23.43}$$

The sum is limited by $2N+2n=1_a+1_b-J-j$. Using again the coefficients and functions listed in appendix 3, we get the following rotational invariants:

$$\langle \text{AO0} \| \text{AA}_\Delta^0 \text{PA}_\Delta^0 \| \text{AO0} \rangle^0 = (8\pi^2/\theta_{\alpha\beta}^2) \cdot \exp(-\xi_{\alpha\beta}^2 \text{AA}_\Delta^2) \cdot F_0(\theta_{\alpha\beta}^2 \text{PA}_\Delta^2)$$

$$\langle \text{AO1} \| \text{AA}_\Delta^1 \text{PA}_\Delta^0 \| \text{AO0} \rangle^1 = -(16\pi^2 \beta \xi_{\alpha\beta}/\theta_{\alpha\beta}^3) \cdot \exp(-\xi_{\alpha\beta}^2 \text{AA}_\Delta^2) \cdot F_0(\theta_{\alpha\beta}^2 \text{PA}_\Delta^2)$$

$$\langle \text{AO1} \| \text{AA}_\Delta^0 \text{PA}_\Delta^1 \| \text{AO0} \rangle^1 = -(16\pi^2 \beta/\theta_{\alpha\beta}^2) \cdot \exp(-\xi_{\alpha\beta}^2 \text{AA}_\Delta^2) \cdot F_1(\theta_{\alpha\beta}^2 \text{PA}_\Delta^2)$$

$$\langle \text{AO1} \| \text{AA}_\Delta^1 \text{PA}_\Delta^1 \| \text{AO1} \rangle^0 = (32\pi^2 (\beta^2 - \alpha^2) \xi_{\alpha\beta}/\theta_{\alpha\beta}) \exp(-\xi_{\alpha\beta}^2 \text{AA}_\Delta^2) \cdot F_1(\theta_{\alpha\beta}^2 \text{PA}_\Delta^2)$$

$$\langle \text{AO1} \| \text{AA}_\Delta^1 \text{PA}_\Delta^1 \| \text{AO1} \rangle^1 = (32\pi^2 \xi_{\alpha\beta}/\theta_{\alpha\beta}) \cdot \exp(-\xi_{\alpha\beta}^2 \text{AA}_\Delta^2) \cdot F_1(\theta_{\alpha\beta}^2 \text{PA}_\Delta^2) \tag{23.44}$$

$$\langle \text{AO1} \| \text{AA}_\Delta^1 \text{PA}_\Delta^1 \| \text{AO1} \rangle^2 = (32\pi^2 (\beta^2 - \alpha^2) \xi_{\alpha\beta}/\theta_{\alpha\beta}^3) \cdot \exp(-\xi_{\alpha\beta}^2 \text{AA}_\Delta^2) \cdot F_1(\theta_{\alpha\beta}^2 \text{PA}_\Delta^2)$$

$$\langle \text{AO1} \| \text{AA}_\Delta^2 \text{PA}_\Delta^0 \| \text{AO1} \rangle^2 = (8\sqrt{30}\pi^2 \xi_{\alpha\beta}^3/5\theta_{\alpha\beta}^3) \cdot \exp(-\xi_{\alpha\beta}^2 \text{AA}_\Delta^2) \cdot F_0(\theta_{\alpha\beta}^2 \text{PA}_\Delta^2)$$

$$\langle \text{AO1} \| \text{AA}_\Delta^0 \text{PA}_\Delta^2 \| \text{AO1} \rangle^2 = (32\sqrt{30}\pi^2 \alpha\beta/5\theta_{\alpha\beta}^2) \cdot \exp(-\xi_{\alpha\beta}^2 \text{AA}_\Delta^2) \cdot F_2(\theta_{\alpha\beta}^2 \text{PA}_\Delta^2)$$

The fifth invariant is the one needed in the example (23.41).

Appendix 1: Projection operators

In eq. (5.1) we have expressed the s.-a. LCAOs by the general SALC coefficients (5.2). This representation is more expedient than that by projection operators. But, of course, it is possible to interrelate the s.-a. functions generated by both methods.

The projection operators or more generally the shift operators of an irreducible representation c are defined by

$$P^c_{ik} = (\dim c/\text{ord}G)\sum_{g\epsilon G} D^c_{ik}(g)^* \cdot U(g) \tag{A1.1}$$

If we apply such an operator to the atomic function $|Aj\varphi_a ap_a\rangle$ defined by (4.3), the resultant transforms according to the representation c, component i. The indices k, j, and p_a are redundant; but an unfortunate choice of them can yield zero, although a s.-a. function of species c does exist. Since the orbitals $|Aj\varphi_a ap_a\rangle$ induce the product representation $\sigma^A \times a$ according to (5.3), the decomposition of the representation obeys the character rule resulting from (2.10):

$$n(Aa,c) = (1/\text{ord}G)\sum_C \mathcal{K}(C)\chi^c(C)^*\chi^a(C)\sigma^A(C) \tag{A1.2}$$

If now n(Aa,c) > 1, there are several independent sets of species c, which may be found by trying several combinations of the indices k, j, and p_a. On the contrary, the indices ε, e, and γ in (5.1) exactly exhaust the set of species c induced by $|Aj\varphi_a ap_a\rangle$.

We calculate the result of the projection. Because of (5.3) we get

$$P^c_{1k}|Aj\varphi_a am_a\rangle = (\dim c/\text{ord}G)\sum_{g\epsilon G} \sum_{n_a 1} D^c_{1k}(g)^* D^a_{n_a m_a}(g)\sigma^A_{ij}(g)|Ai\varphi_a an_a\rangle \tag{A1.3}$$

If we apply P^c_{1k} to the molecular orbitals (5.1), we derive the following interrelation:

$$|(Ae\varepsilon,\varphi_a a)\gamma cl\rangle = P^c_{1k}|(Ae\varepsilon,\varphi_a a)\gamma ck\rangle = \sum_{jr_a} K(\gamma ck, Aj\varepsilon e, ar_a)P^c_{1k}|Ai\varphi_a ar_a\rangle \tag{A1.4}$$

This shows that the s.-a. MOs determined by (5.1) may be regarded as linear combinations of the projected orbitals (A1.3), where the sum is taken for the redundant indices j and r_a. The index k is arbitrary, but has to be the same in the SALC coefficient and the projector.

The orthogonality relations of the SALC coefficients resulting from (2.26/27) and (3.6/7) allow the inversion of (A1.4). Starting from $P^c_{1m}|(Ae\varepsilon,\varphi_a a)\gamma ck\rangle = \delta(m,k)\cdot|(Ae\varepsilon,\varphi_a a)\gamma cl\rangle$ one derives:

$$P^c_{1m}|Aj\varphi_a as_a\rangle = \sum_{\varepsilon e\gamma} K(\gamma cm, Aj\varepsilon e, as_a)\cdot|(Ae\varepsilon,\varphi_a a)\gamma cl\rangle \tag{A1.5}$$

This completes the wanted interrelations.

Appendix 2: Molecular integrals of Gaussian functions

In favor of the general validity, no reference was made to special orbital systems in section 13. Only the choice of the distance vectors between the atomic centres has regard for the integrals of Gaussian or related orbitals. This appendix has two aims: At first the theorems, i.e. their rotational invariants, are illustrated for a particular system of orbitals; and secondly the proofs of the strong theorems concerning the three- and four-centre integrals are supplemented using this special system of orbitals.

As well known, the GTOs are recommended by their simple integrals. We therefore choose the radial functions in (13.1) from this type. As explained in section 13, only spherical orbitals have the adequate tensorial structure. Such orbital systems have been discussed in [45-47]. The most suitable for our purpose is that of [46], and we routinely refer to this paper. The advantage of this system results from its generation by the gradient operator:

$$\langle \vec{r} | \alpha n l m \rangle = \alpha^{-2n-1} \Delta^n \mathcal{Y}_{lm}(\vec{\nabla}) \exp(-\alpha^2 r^2) \qquad (A2.1)$$

The right transformation property is achieved by inserting the operator into the solid harmonics $\mathcal{Y}_{lm}(\vec{a}) = (ia)^l Y_{lm}(\vec{a}/a)$. As a function of operators we distinguish $\mathcal{Y}_{lm}(\vec{a})$ from $\langle \vec{a} | sol \, lm \rangle$ in order to avoid confusion.

Emphasizing the general validity of the theorems, we sketch the relations of this system to other orbital systems. As shown in [46], the definition (A2.1) leads to the following radial functions:

$$N_1(nl) \cdot L_n^{l+1/2}(\alpha^2 r^2) \cdot \exp(-\alpha^2 r^2)$$

Therefore the ordinary spherical GTOs as well as the spherical oscillator functions can be expanded as

$$(\alpha r)^{2n} \exp(-\alpha^2 r^2) \langle \alpha \vec{r} | sol \, lm \rangle = N_2(nl) \cdot \sum_{k=0}^{n} \binom{n+l+1/2}{n-k} (-1)^k \langle \vec{r} | \alpha k l m \rangle \quad (A2.2)$$

$$\langle \vec{r} | Osc \, \alpha n l m \rangle = N_3(nl) \cdot \sum_{k=0}^{n} \binom{n+l+1/2}{n-k} 2^k \langle \vec{r} | \alpha k l m \rangle , \qquad (A2.3)$$

where $N_i(nl)$ are normalization constants. For other systems of orbitals, $\langle \vec{r} | X n l m \rangle = R_{nl}^X(r) \langle \vec{r} | sol \, lm \rangle$, result infinite series expansions with respect to the radial quantum number. The expansion coefficients can be calculated, because the polynoms

$$N_4(nl) \cdot L_n^{l+1/2}(\alpha^2 r^2) \langle sol \, lm | \vec{r} \rangle \qquad (A2.4)$$

supplement the system (A2.1) to a biorthonormal system. Thence the coefficients of the expansion

$$\langle \vec{r} | X n l m \rangle = \sum_k c_k^{Xn} \langle \vec{r} | \alpha n l m \rangle \qquad (A2.5)$$

are given by:

$$c_k^{Xn} = N_4(nl)\int_0^\infty L_k^{l+1/2}(\alpha^2 r^2)R_{nl}^X(r)r^{2l+2}dr \tag{A2.6}$$

The multi-centre integrals contain products of the functions (A2.1) with respect to different centres. These products can be reorganized by a theorem given in [46], eq.(3.3):

$$\alpha_1^{-2n_1-1}\Delta_1^{n_1}\mathscr{Y}_{l_1m_1}(\vec{\nabla}_1)\cdot\alpha_2^{-2n_2-1}\Delta_2^{n_2}\mathscr{Y}_{l_2m_2}(\vec{\nabla}_2)$$

$$= \sum_{n_3 l_3 n_4 l_4 L}\sum [n_3 l_3 n_4 l_4 L \| \varphi_{12} \| n_1 l_1 n_2 l_2 L]\begin{pmatrix}l_1 l_2 L \\ m_1 m_2 M\end{pmatrix}^{+}\cdot\begin{pmatrix}l_3 l_4 L \\ m_3 m_4 M\end{pmatrix} \tag{A2.7}$$

$$\cdot \xi_{12}^{-2n_3-1_3}\Delta_3^{n_3}\mathscr{Y}_{l_3m_3}(\vec{\nabla}_3)\cdot\Theta_{12}^{-2n_4-1_4}\Delta_4^{n_4}\mathscr{Y}_{l_4m_4}(\vec{\nabla}_4)$$

where $\vec{\nabla}_3=\vec{\nabla}_1-\vec{\nabla}_2$, $\vec{\nabla}_4=(\alpha_1^2\vec{\nabla}_1+\alpha_2^2\vec{\nabla}_2)/\Theta_{12}$, $\varphi_{12}=\tan^{-1}(\alpha_1/\alpha_2)$, $\Theta_{12}=\sqrt{\alpha_1^2+\alpha_2^2}$, and $\xi_{12}=\alpha_1\alpha_2/\Theta_{12}$. The sums in (A2.7) are limited by $2n_3+l_3+2n_4+l_4=2n_1+l_1$ $+2n_2+l_2$. As for the coefficients of this expansion we again refer to [45, 46], and to appendix 3.

The application of the operator $\mathscr{Y}_{lm}(\vec{\nabla})$ to a scalar function $f(r)$ results in

$$\mathscr{Y}_{lm}(\vec{\nabla})f(r) = g^l(r)\langle\vec{r}|\text{sol } lm\rangle, \tag{A2.8}$$

where the scalar functions are given by $g^l(r)=\sqrt{1/4\pi}\cdot(\frac{1}{r}\cdot\frac{d}{dr})^l f(r)$. Because of (A2.7), we need the corresponding theorem with respect to the scalar functions of two vector variables $\vec{r}_1$ and $\vec{r}_2$. This is given by

$$\mathscr{Y}_{l_1m_1}(\vec{\nabla}_1)\mathscr{Y}_{l_2m_2}(\vec{\nabla}_2)f(r_1^2,r_2^2,\vec{r}_1\cdot\vec{r}_2)$$

$$= \sum_{L\lambda_1\lambda_2}g_{\lambda_1\lambda_2}^{Ll_1l_2}(r_1^2,r_2^2,\vec{r}_1\cdot\vec{r}_2)\sqrt{2L+1}^{-1}\begin{pmatrix}l_1 l_2 L \\ m_1 m_2 M\end{pmatrix}^{+}\langle\vec{r}_1,\vec{r}_2|\text{sol}(\lambda_1\lambda_2)LM\rangle, \tag{A2.9}$$

where the sums are limited by $\lambda_1+\lambda_2 \leqslant l_1+l_2$ and $l_1+l_2-\lambda_1-\lambda_2=$ even. An infinite expansion of this type is trivial from the tensorial point of view. The limitation of the sums is the essential statement of (A2.9).

The proof is given by a recursive calculation of the scalar functions $g_{\lambda_1\lambda_2}^{Ll_1l_2}$. (A2.9) is obviously valid, if $l_1=l_2=0$. The first function then is $g_{oo}^{obo}=f$. If now (A2.9) and the limitation hold for a pair of angular momenta l_1 and l_2, we deduce the same for the pair l_1+1 and l_2. For this purpose, we start from $\mathscr{Y}_{l_1+1m_1}(\vec{\nabla}_1)\mathscr{Y}_{l_2m_2}(\vec{\nabla}_2)f(...)$ and decompose the first spherical harmonic by:

$$\mathscr{Y}_{l_1+1m_1}(\vec{\nabla}_1) = (2l_1+3)\sqrt{4\pi/3(l_1+1)}\sum\begin{pmatrix}1^{+} 1^{+}l_1+1 \\ m_1 \mu m_1\end{pmatrix}\mathscr{Y}_{l_1m_1}(\vec{\nabla}_1)\mathscr{Y}_{1\mu}(\vec{\nabla}_1)$$

In order to perform the operation $\mathscr{Y}_{1\mu}(\vec{\nabla})=\sqrt{3/4\pi}\nabla_\mu$, we define the following derivatives:

$$g_{\lambda_1\lambda_2 k}^{Ll_1l_2}(x_1,x_2,x_3) = \frac{\partial}{\partial x_k}g_{\lambda_1\lambda_2}^{Ll_1l_2}(x_1,x_2,x_3)$$

This yields:

$$\mathscr{Y}_{1\mu}(\vec{\nabla}_1)g_{\lambda_1\lambda_2}^{Ll_1l_2}(r_1^2,r_2^2,\vec{r}_1\cdot\vec{r}_2) =$$

$$= 2g_{\lambda_1\lambda_2 1}^{L1_1 1_2}(r_1^2,r_2^2,\vec{r}_1\cdot\vec{r}_2)\langle\vec{r}_1|sol\ 1\mu\rangle + g_{\lambda_1\lambda_2 3}^{L1_1 1_2}(r_1^2,r_2^2,\vec{r}_1\cdot\vec{r}_2)\langle\vec{r}_2|sol\ 1\mu\rangle$$

and

$$\mathcal{Y}_{1\mu}(\vec{\nabla}_1)\langle\vec{r}_1|sol\ \lambda_1\mu_1\rangle = (2\lambda_1+1)\sqrt{3\lambda_1/4\pi}\cdot\sum\binom{\lambda_1-1\ +1\ \lambda_1}{\mu_1'\quad\mu\quad\mu_1}\langle\vec{r}_1|sol\ \lambda_1-1\mu_1'\rangle$$

If we now insert these expressions into $\mathcal{Y}_{1_1+1m_1}(\vec{\nabla}_1)\mathcal{Y}_{1_2m_2}(\vec{\nabla}_2)f(\dots)$, we get a new equation of type (A2.9) with l_1+1 instead of l_1. The comparison of the coefficients of the functions $\langle\vec{r}_1,\vec{r}_2|sol(\lambda_1\lambda_2)LM\rangle$ then yields the recursive formula of the scalar functions:

$$g_{j_1 j_2}^{Jl_1+1l_2} = -(2l_1+3)\sqrt{4\pi/3(l_1+1)}\sum_L(2L+1)\begin{Bmatrix}l_1 l_2 & L\\ J\ 1 & l_1+1\end{Bmatrix}\cdot\Big[\sum_{\lambda_1}\begin{Bmatrix}J & j_1 j_2\\ \lambda_1 L\ 1\end{Bmatrix}\langle j_1\|1\|\lambda_1\rangle$$

$$\cdot(2g_{\lambda_1 j_2 1}^{Ll_1 l_2}\cdot r_1^{\lambda_1+1-j_1} + g_{\lambda_1 j_2}^{Ll_1 l_2}\cdot(2\lambda_1+1)\delta(j_1,\lambda_1-1))\qquad\text{(A2.10)}$$

$$+(-1)^{J+L+1}\sum_{\lambda_2}\begin{Bmatrix}J & j_1 j_2\\ \lambda_2 1\ L\end{Bmatrix}\langle j_2\|1\|\lambda_2\rangle g_{j_1\lambda_2 3}^{Ll_1 l_2}\cdot r_2^{\lambda_2+1-j_2}\Big]$$

Since $g_{\lambda_1 j_2}^{Ll_1 l_2}\neq 0$ only, if $\lambda_1+j_2\leq l_1+l_2$, and $\langle j_1\|1\|\lambda_1\rangle\neq 0$ only, if $j_1\leq\lambda_1+1$, we have the limitation $j_1+j_2\leq\lambda_1+1+j_2\leq l_1+l_2+1$. As for the second summand, we conclude in the same way $j_1+\lambda_2\leq l_1+l_2$, $j_2\leq\lambda_2+1$ and therefore $j_1+j_2\leq j_1+\lambda_2+1\leq l_1+l_2+1$. From both limits follows that $g_{j_1 j_2}^{Ll_1+1l_2}\neq 0$ only, if $j_1+j_2\leq l_1+l_2+1$, which had to be shown. The recursion with respect to l_2 is found by a fitting exchange of indices. In the long run it is, of course, desirable to derive an explicit formula of the functions g.

The analogue of (A2.9) involving three vector variables $\vec{r}_1$, $\vec{r}_2$, and $\vec{r}_3$ is given by:

$$\mathcal{Y}_{1_1 m_1}(\vec{\nabla}_1)\mathcal{Y}_{1_2 m_2}(\vec{\nabla}_2)\mathcal{Y}_{1_3 m_3}(\vec{\nabla}_3)f(r_1^2,r_2^2,r_3^2,\vec{r}_1\cdot\vec{r}_2,\vec{r}_2\cdot\vec{r}_3,\vec{r}_3\cdot\vec{r}_1)$$

$$= \sum_{j_1 JL}\sum_{\lambda}h\begin{Bmatrix}1_1 1_2\\ j_1 j_2\end{Bmatrix}_{Jj_3}^{L1_3}(r_1^2,r_2^2,r_3^2,\vec{r}_1\cdot\vec{r}_2,\vec{r}_2\cdot\vec{r}_3,\vec{r}_3\cdot\vec{r}_1)\qquad\text{(A2.11)}$$

$$\cdot\binom{1_1 1_2 L}{m_1 m_2 M}\binom{L\ 1_3\ \lambda}{M\ m_3\mu}\langle\vec{r}_1,\vec{r}_2,\vec{r}_3|sol(j_1 j_2)Jj_3\lambda\mu\rangle$$

The essential limitation of the sums is now $j_1+j_2+j_3\leq l_1+l_2+l_3$. The scalar functions h again can be calculated recursively.

We now are prepared to discuss the particular integrals. As an example of the two-centre integrals (13.3), we choose those involving the momentum operator. For this case, in [46], the integral formula (4.5) has been derived, which reads:

$$\langle A\alpha n_a l_a m_a|\nabla_\mu|B\beta n_b l_b m_b\rangle\qquad\text{(A2.12)}$$

$$= \sum_{LI}\mathcal{L}_{\alpha\beta}^0(n_a n_b,Ll_a l_b,AB)\binom{l+1}{m_a}\binom{L^+}{m}\binom{1+1}{m}\binom{1_b}{\mu\ m_b}\langle\zeta_{\alpha\beta}\vec{AB}|sol\ LM\rangle,$$

where $\zeta_{\alpha\beta}$ has been defined above and $\mathcal{L}_{\alpha\beta}^0(\dots)$ is given by eq.(4.6) of [46]. From this follows the generalized reduced matrix element. Because of the symmetric coupling used in (13.3), a 6j symbol is involved:

$$\langle n_a l_a \| AB_j^J \nabla \| n_b l_b \rangle^J = \sqrt{(2J+1)/4\pi} \cdot \sum_I (-1)^{1+l_a+l_b} \begin{Bmatrix} 1 & a^J & 1 \\ 1 & a^1 & j & b \end{Bmatrix} \cdot \xi_{\alpha\beta}^j \eta_{\alpha\beta}^{10} (n_a n_b, j1 a_l b_1, AB) \tag{A2.13}$$

An example of the weak theorem of the three-centre integrals (13.8) is given by eq.(5.7) of [46]. The rotational invariants now are:

$$\langle A n_a l_a \| AB^J PC^j \| B n_b l_b \rangle^L = (4\pi/(2L+1)) \xi_{\alpha\beta}^J \theta_{\alpha\beta}^j \eta_{\alpha\beta}^0 (n_a n_b, L l_a l_b, Jj, AB.PC) \tag{A2.14}$$

According to [46], eq.(5.8), they depend on the distances AB and PC only. From the same equation follows the limitation of the sums in (13.8) by $J+j \leqslant 2n_a+l_a+2n_b+l_b$. Therefore $J+j$ may be greater than l_a+l_b.

In order to prove the strong theorem we, start from an integral over two ns-orbitals, which is an invariant by itself:

$$\langle A n_a 00 | r_C^{-1} | B \beta n_b 00 \rangle = \sum_J \eta_{\alpha\beta}^0 (n_a n_b, 000, JJ, AB, PC)(\alpha\beta AB \cdot PC)^J P_J \left(\frac{\overrightarrow{AB \cdot PC}}{AB \cdot PC} \right) \tag{A2.15}$$

where P_J is a Legendre polynomial. In accordance to (A2.1), the higher functions are generated by $|\alpha n l m\rangle = \alpha^{-1} \mathscr{Y}_{lm}(\nabla) |\alpha n 00\rangle$ and we can represent the shifted orbitals by $|A \alpha n_a l_a m_a\rangle = (-\alpha)^{-1 a} \mathscr{Y}_{l_a m_a}(\nabla_A) |A \alpha n_a 00\rangle$. Since the gradient now operates on the parameters $\overrightarrow{A}$, we can interchange it with the integration and generate the higher integrals from (A2.15) by:

$$\langle A \alpha n_a l_a m_a | r_C^{-1} | B \beta n_b l_b m_b \rangle \tag{A2.16}$$

$$= (-\alpha)^{-1a} (-\beta)^{-1b} \mathscr{Y}_{l_a m_a}(\overrightarrow{\nabla}_A) \mathscr{Y}_{l_b m_b}(\overrightarrow{\nabla}_B) \langle A \alpha n_a 00 | r_C^{-1} | B \beta n_b 00 \rangle$$

$$= \alpha^{-1a} \cdot \beta^{-1b} \cdot \mathscr{Y}_{l_a m_a}(\overrightarrow{\nabla}_{AC}) \mathscr{Y}_{l_b m_b}(\overrightarrow{\nabla}_{BC}) \langle A \alpha n_a 00 | r_C^{-1} | B \beta n_b 00 \rangle$$

If we regard the integral $\langle A \alpha n_a 00 | r_C^{-1} | B \beta n_b 00 \rangle$ as a function of $\overrightarrow{AC}$ and $\overrightarrow{BC}$ (which is possible), the theorem (A2.9) immediately supplies the intended result (13.9). But because of the more natural interpretation as a function of $\overrightarrow{AB}$ and $\overrightarrow{PC}$, we reshape the derivations by (A2.7). This yields:

$$\langle A \alpha n_a l_a m_a | r_C^{-1} | B \beta n_b l_b m_b \rangle = \sum_{njNJL} \sum [njNJL \| \varphi_{\alpha\beta} \| 01_a 01_b L] \begin{pmatrix} l+1_b & L+ \\ m_a & m_b & M \end{pmatrix} \begin{pmatrix} j+J+ & L \\ m & M & M \end{pmatrix}$$
$$\cdot \xi_{\alpha\beta}^{-2n-j} \theta_{\alpha\beta}^{-2N-J} \mathscr{Y}_{jm}(\overrightarrow{\nabla}_{AB}) \mathscr{Y}_{JM}(\overrightarrow{\nabla}_{PC}) \Delta_{AB}^n \Delta_{PC}^N \langle A \alpha n_a 00 | r_C^{-1} | B \beta n_b 00 \rangle \tag{A2.17}$$

with $l_a+l_b = 2n+j+2N+J \geqslant j+J$. Because the terms $\Delta_{AB}^n \Delta_{PC}^N \langle A \alpha n_a 00 | r_C^{-1} | B \beta n_b 00 \rangle$ are scalar functions, the theorem (A2.9) applies now with respect to the vectors $\overrightarrow{AB}$ and $\overrightarrow{PC}$. This yields the following expression of the rotational invariants of the strong theorem:

$$\langle A n_a l_a \| AB^{J'} PC^{j'} \| B n_b l_b \rangle^L \tag{A2.18}$$

$$= (1/4\pi(2L+1)^2) \sum_{njNJ} [njNJL \| \varphi_{\alpha\beta} \| 01_a 01_b L] \cdot \xi_{\alpha\beta}^{-j} \theta_{\alpha\beta}^{-J} \cdot g_{n_a n_b J'j'}^{nNjJ}(AB^2, PC^2, \overrightarrow{AB} \cdot \overrightarrow{PC})$$

The scalar functions $g_{n_a n_b J'j'}^{nNjJ}$ have to be calculated recursively by (A2.10) beginning with $g_{n_a n_b 00}^{nNoo} = \xi_{\alpha\beta}^{-2n} \theta_{\alpha\beta}^{-2N} \Delta_{AB}^n \Delta_{PC}^N \langle A \alpha n_a 00 | r_C^{-1} | B \beta n_b 00 \rangle$. From the limit $j+J \geqslant J'+j'$ according to (A2.9) follows $l_a+l_b \geqslant J'+j'$ as proposed.

Since the expansions (A2.2-4) affect the radial quantum numbers only, the rotational invariants of the integrals of other orbital systems have the same group theoretical structure,

$$\langle AXn_a1_a\|AB^J PC^j\|BXn_b1_b\rangle^L = \sum_{n_a'n_b'} c_{n_a}^{n_a'}c_{n_b}^{n_b'} \langle An_a'1_a\|AB^J PC^j\|Bn_b'1_b\rangle^L,$$

and the strong theorem is valid for any system of atomic orbitals.

We finally come to the four-centre integrals of the electronic interaction. First we discuss the weak version of (13.13). In order to separate the functions (13.12), we have chosen the coupling different from that in eq.(6.4) of [46]. Because of the necessary recoupling, the rotational invariants of (13.13) read:

$$[(An_a1_a^+,Cn_c1_c)1\|(AC^{j_4} BD^{j_2})^J_{PQ}{}^{j_3}\|(Bn_b1_b^+,Dn_d1_d)1]^L = \sqrt{1/(4\pi)^3(2l+1)(2l'+1)}$$

$$\cdot \sum_{pq\lambda}\sqrt{2\lambda+1}\cdot\sigma^0(n_an_bn_cn_d,1l_a1_c,l_b1_d,J_3pq,j_1j_2,AC,BD,PQ)\mathcal{E}_{\alpha\gamma}^{j_4}\mathcal{E}_{\beta\delta}^{j_2}\cdot g^{j_3}\begin{Bmatrix} j_4 & j_2 & \lambda \\ p & q & j \\ 1 & 1' & L \end{Bmatrix}$$

with $g=[(\alpha^2+\gamma^2)(\beta^2+\delta^2)/(\alpha^2+\beta^2+\gamma^2+\delta^2)]^{1/2}.$ \qquad (A2.19)

The invariants are functions of the distances AC, BD, and PQ only and not of the angles between them.

The proof of the strong theorem follows the lines of that given for the three-centre integrals and we can be brief. As in (A2.16), one generates the integrals by the gradient operators with respect to the atomic centres:

$$\langle A\alpha n_a1_am_a,B\beta n_b1_bm_b|r_{12}^{-1}|C\gamma n_c1_cm_c,D\delta n_d1_dm_d\rangle=(-\alpha)^{-1_a}(-\beta)^{-1_b}(-\gamma)^{-1_c}(-\delta)^{-1_d}$$

$$\cdot\mathcal{Y}_{1_am_a}(\vec{\nabla}_A)\mathcal{Y}_{1_bm_b}(\vec{\nabla}_B)\mathcal{Y}_{1_cm_c}(\vec{\nabla}_C)\mathcal{Y}_{1_dm_d}(\vec{\nabla}_D)\langle A\alpha n_a00,B\beta n_b00|r_{12}^{-1}|C\gamma n_c00,D\delta n_d00\rangle$$

By repeated application of (A2.7), the differential operators are adapted to the distances $\vec{AC}$, $\vec{BD}$, and $\vec{PQ}$. This leads to the scalar functions

$$f(AC^2,BD^2,PQ^2,\vec{AC}\cdot\vec{BD},\vec{BD}\cdot\vec{PQ},\vec{PQ}\cdot\vec{AC}) = \xi_{\alpha\gamma}^{-2n_a}\xi_{\beta\delta}^{-2n'}\xi_{\sigma\tau}^{-2N-2N'-J-J'+K}$$

$$\cdot\Delta_{AC}^n\Delta_{BD}^{n'}\Delta_{PQ}^{N+N'+(J+J-K)/2}\langle A\alpha n_a00,B\beta n_b00|r_{12}^{-1}|C\gamma n_c00,D\delta n_d00\rangle$$

with $\sigma=\xi_{\alpha\gamma}$ and $\tau=\xi_{\beta\delta}$. Because of (A2.11) the derivatives of these functions have the structure

$$\mathcal{Y}_{Jm}(\vec{\nabla}_{AC})\mathcal{Y}_{J'm'}(\vec{\nabla}_{BD})\mathcal{Y}_{KM}(\vec{\nabla}_{PQ})f$$

$$= \sum_{J_1IkL}h\begin{Bmatrix}jj'\\j_1j_2\end{Bmatrix}^{kKL}_{Ij_3}(AC^2,\ldots,\vec{PQ}\cdot\vec{AC})\cdot\begin{pmatrix}jj'k\\mm'\mu\end{pmatrix}^+\begin{pmatrix}kKL\\\mu MM\end{pmatrix}^+\langle\vec{AC},\vec{BD},\vec{PQ}|sol(j_1j_2)Ij_3LM\rangle$$

with the limitation $j_1+j_2+j_3 \leqslant j+j'+K$. In analogy to (A2.18), this leads to the rotational invariants of the strong theorem (13.13). Using the abbreviation $F = 1/\sqrt{(4\pi)^3(2l+1)(2l'+1)(2L+1)(2K+1)}$, the result reads:

$$[(An_a1_a^+,Cn_c1_c)1\|(AC^{j_4} BD^{j_2})^I_{PQ}{}^{j_3}\|(Bn_b1_b^+,Dn_d1_d)1]^L = \qquad (A2.20)$$

$$= F \cdot \sum_{njNJ} \sum_{n'j'NJ'} \sum_{kK} [njNJl\|\varphi_{\alpha\gamma}\|01_a01_cl] \cdot [n'j'NJl\|\varphi_{\beta\delta}\|01_b01_dl]$$

$$\cdot [N+N'+(J+J'-K)/2, KOOK\|\varphi_{\sigma\tau}\|NJNJK] \begin{Bmatrix} j & J & l \\ j' & J' & l' \\ k & K & L \end{Bmatrix} \cdot \zeta_{\alpha\gamma}^{-j} \zeta_{\beta\delta}^{-j} \zeta_{\sigma\tau}^{-K} \qquad (A2.20)$$

$$\cdot h \begin{Bmatrix} j\ j')kKL \\ j_1 j_2)Ij_3 \end{Bmatrix} (AC^2, BD^2, PQ^2, \vec{AC} \cdot \vec{BD}, \vec{BD} \cdot \vec{PQ}, \vec{PQ} \cdot \vec{AC})$$

The balance of the angular momenta follows from $l_a + l_c = 2n + j + 2N + J \geqslant j + J$, $l_b + l_d = 2n' + j' + 2N' + J' \geqslant j' + J'$, and $J + J' \geqslant K$. This results in $l_a + l_b + l_c + l_d \geqslant j + j' + J + J' \geqslant j + j' + K \geqslant j_1 + j_2 + j_3$, which had to be shown.

Appendix 3: Modified Moshinsky-Smirnov coefficients and Gauss-Laguerre type functions

In this appendix, we compile the coefficients involved in the eqs. (13.10/11), (23.21 and 43), (A2.7, 17/18 and 20) as far as they are neeeded in section 23. The coefficients are related to the Moshinsky-Smirnov coefficients by a different normalization:

$$[n_1l_1,n_2l_2,L\|\varphi\|n_3l_3,n_4l_4,L] \tag{A3.1}$$

$$= (2L+1)\cdot A(n_1l_1)A(n_2l_2)A(n_3l_3)^{-1}A(n_4l_4)^{-1}\langle n_1l_1,n_2l_2L\|\varphi\|n_3l_3,n_4l_4,L\rangle,$$

where

$$A(nl) = \sqrt{4\pi/(2n+2l+1)!!(2n)!!} \tag{A3.2}$$

The Moshinsky-Smirnov coefficients have been calculated by the formulae (11) and (23) in a paper of Trlifaj [72]. In addition to the references given in [45], we mention the papers of Dobeš [73] and Niukkanen [74]. The coefficients are arranged according to the sum $K=2n_1+l_1+2n_2+l_2=2n_3+l_3+2n_4+l_4$ and to the angular momentum quantum number L.

Table 29. The modified Moshinsky-Smirnov coefficients

K L	$n_1l_1n_2l_2$	$n_3l_3n_4l_4$	$[n_1l_1,n_2l_2,L\|\varphi\|n_3l_3,n_4l_4,L]$
0 0	0 0 0 0	0 0 0 0	1
1 1	0 1 0 0	0 1 0 0	$3\cos\varphi$
	0 1 0 0	0 0 0 1	$-3\sin\varphi$
	0 0 0 1	0 1 0 0	$3\sin\varphi$
	0 0 0 1	0 0 0 1	$3\cos\varphi$
2 0	1 0 0 0	1 0 0 0	$\cos^2\varphi$
	1 0 0 0	0 0 1 0	$\sin^2\varphi$
	1 0 0 0	0 1 0 1	$-\sqrt{3}\sin\varphi\cdot\cos\varphi$
	0 0 1 0	1 0 0 0	$\sin^2\varphi$
	0 0 1 0	0 0 1 0	$\cos^2\varphi$
	0 0 1 0	0 1 0 1	$\sqrt{3}\sin\varphi\cdot\cos\varphi$
	0 1 0 1	1 0 0 0	$\sqrt{3}\sin\varphi\cdot\cos\varphi$
	0 1 0 1	0 0 1 0	$-\sqrt{3}\sin\varphi\cdot\cos\varphi$
	0 1 0 1	0 1 0 1	$\cos^2\varphi-\sin^2\varphi$
2 1	0 1 0 1	0 1 0 1	3
2 2	0 2 0 0	0 2 0 0	$5\cos^2\varphi$
	0 2 0 0	0 0 0 2	$5\sin^2\varphi$
	0 2 0 0	0 1 0 1	$-\sqrt{30}\sin\varphi\cdot\cos\varphi$
	0 0 0 2	0 2 0 0	$5\sin^2\varphi$
	0 0 0 2	0 0 0 2	$5\cos^2\varphi$
	0 0 0 2	0 1 0 1	$\sqrt{30}\sin\varphi\cdot\cos\varphi$
	0 1 0 1	0 2 0 0	$\sqrt{30}\sin\varphi\cdot\cos\varphi$
	0 1 0 1	0 0 0 2	$-\sqrt{30}\sin\varphi\cdot\cos\varphi$

Table 29. (continued)

K L	$n_1l_1n_2l_2$	$n_3l_3n_4l_4$	$[n_1l_1,n_2l_2,L\|\varphi\|n_3l_3,n_4l_4,L]$
2 2	0 1 0 1	0 1 0 1	$5(\cos^2\varphi-\sin^2\varphi)$
3 1	1 1 0 0	1 1 0 0	$3\cos^3\varphi$
	1 1 0 0	1 0 0 1	$3\sin\varphi\cdot\cos^2\varphi$
	1 1 0 0	0 1 1 0	$3\sin^2\varphi\cdot\cos\varphi$
	1 1 0 0	0 0 1 1	$-3\sin^3\varphi$
	1 1 0 0	0 2 0 1	$-3\sqrt{2}\sin^2\varphi\cdot\cos\varphi$
	1 1 0 0	0 1 0 2	$3\sqrt{2}\sin\varphi\cdot\cos^2\varphi$
	0 0 1 1	1 1 0 0	$3\sin^3\varphi$
	0 0 1 1	1 0 0 1	$3\sin^2\varphi\cdot\cos\varphi$
	0 0 1 1	0 1 1 0	$3\sin\varphi\cdot\cos^2\varphi$
	0 0 1 1	0 0 1 1	$3\cos^3\varphi$
	0 0 1 1	0 2 0 1	$3\sqrt{2}\sin^2\varphi\cdot\cos\varphi$
	0 0 1 1	0 1 0 2	$3\sqrt{2}\sin\varphi\cdot\cos^2\varphi$
4 0	2 0 0 0	2 0 0 0	$\cos^4\varphi$
	2 0 0 0	1 0 1 0	$\sin^2\varphi\cdot\cos^2\varphi$
	2 0 0 0	0 0 2 0	$\sin^4\varphi$
	2 0 0 0	1 1 0 1	$-\sqrt{3}\sin\varphi\cdot\cos^3\varphi$
	2 0 0 0	0 1 1 1	$\sqrt{3}\sin^3\varphi\cdot\cos\varphi$
	2 0 0 0	0 2 0 2	$\sqrt{5}\sin^2\varphi\cdot\cos^2\varphi$
4 2	1 2 0 0	0 1 1 1	$-\sqrt{30}\sin^3\varphi\cdot\cos\varphi$
	0 0 1 2	0 1 1 1	$\sqrt{30}\sin\varphi\cdot\cos^3\varphi$
	0 0 1 2	1 1 0 1	$\sqrt{30}\sin^3\varphi\cdot\cos\varphi$

For the calculations in section 23, we further list the Gauss-Laguerre type functions occurring in the eqs.(23.21 and 43):

$$\varphi^0(\alpha,-1l,r) = (-2)^{1-1}F_1(\alpha^2 r^2)$$

$$\varphi^0(\alpha,\ 0l,r) = (-2)^{1}\cdot\exp(-\alpha^2 r^2)$$

$$\varphi^0(\alpha,\ 1l,r) = (-2)^{1+1}(2l+3-2\alpha^2 r^2)\cdot\exp(-\alpha^2 r^2)$$

$$\varphi^0(\alpha,2l,r) = (-2)^{1+2}\left[(2l+3)(2l+5)-4(2l+5)\alpha^2 r^2+4\alpha^4 r^4\right]\cdot\exp(-\alpha^2 r^2)$$

Appendix 4. Isoscalar factors and s.-a. solid harmonics of the group T_d

In the following, the indices g (gerade) and u (ungerade) mark the even or odd parity of the irreducible representations of the group SO(3).

Table 30. SO(3) compatibility table for the group T_d.

SO(3)	0_g	0_u	1_g	1_u	2_g	2_u
T_d	A_1	A_2	T_1	T_2	$E+T_2$	$E+T_1$

Table 31. The isoscalar factors $Is(\begin{smallmatrix}jkl\\abc\end{smallmatrix})$ with $j+k+l\leqslant 4$.

j	k	l	a	b	c	Is		j	k	l	a	b	c	Is
0_g	0_g	0_g	A_1	A_1	A_1	1		2_g	2_g	0_g	E	E	A_1	$\sqrt{2/5}$
0_u	0_u	0_g	A_2	A_2	A_1	1		2_g	2_g	0_g	T_2	T_2	A_1	$\sqrt{3/5}$
								2_u	2_u	0_g	E	E	A_1	$\sqrt{2/5}$
1_g	1_g	0_g	T_1	T_1	A_1	1		2_u	2_u	0_g	T_1	T_1	A_1	$\sqrt{3/5}$
1_u	1_u	0_g	T_2	T_2	A_1	1		2_g	2_u	0_u	E	E	A_2	$\sqrt{2/5}$
1_g	1_u	0_u	T_1	T_2	A_2	1		2_g	2_u	0_u	T_2	T_1	A_2	$\sqrt{3/5}$
1_g	1_g	1_g	T_1	T_1	T_1	1								
1_u	1_u	1_g	T_2	T_2	T_1	1								
1_g	1_g	2_g	T_1	T_1	E	$\sqrt{2/5}$								
1_g	1_g	2_g	T_1	T_1	T_2	$\sqrt{3/5}$								
1_u	1_u	2_g	T_2	T_2	E	$\sqrt{2/5}$								
1_u	1_u	2_g	T_2	T_2	T_2	$\sqrt{3/5}$								
1_g	1_u	2_u	T_1	T_2	E	$\sqrt{2/5}$								
1_g	1_u	2_u	T_1	T_2	T_1	$\sqrt{3/5}$								

Table 32. The s.-a. solid harmonics of the group T_d

l	a	p	$\langle \vec{r}|$sol lap$\rangle$
0	A_1	1	$1/\sqrt{4\pi}$
1	T_2	x	$i\sqrt{3/4\pi}\cdot x$
1	T_2	y	$i\sqrt{3/4\pi}\cdot y$
1	T_2	z	$i\sqrt{3/4\pi}\cdot z$
2	E	1	$-\sqrt{5/16\pi}(3z^2-r^2)$
2	E	2	$-\sqrt{15/16\pi}(x^2-y^2)$
2	T_2	x	$-\sqrt{15/4\pi}\cdot yz$
2	T_2	y	$-\sqrt{15/4\pi}\cdot zx$
2	T_2	z	$-\sqrt{15/4\pi}\cdot xy$

References

1. W.Heisenberg: Schritte über Grenzen (page 23).
    Piper, München, 1971
2. W.Heisenberg: Platon und die moderne Physik. quoted from E.Hunger:
    Von Demokrit bis Heisenberg (page 96). Vieweg, Braunschweig,1964
3. L.Pauling and R.Hayward: The Architecture of Molecules.Freeman,
    San Francisco, 1964
4. B.G.Wybourne: The "Gruppenpest" Yesterday, Today, and Tomorrow.
    Int.J.Quantum Chem.Symp. 7,35 (1973)
5. J.S.Griffith: The Irreducible Tensor Method for Molecular Symmetry
    Groups. Prentice-Hall, Englewood Cliffs (N.J.), 1962
6. B.L.Silver: Irreducible Tensor Methods.Academic Press, New York,1976
7. C.D.H.Chisholm: Group Theoretical Techniques in Quantum Chemistry.
    Academic Press, London, 1976
8. W.Haberditzl: Quantenchemie, Vol. 4, Komplexverbindungen.
    Hüthig, Heidelberg, 1979
9. E.König and S.Kremer: Ligand Field Energy Diagrams. Plenum Press,
    New York, 1977
10. G.Fieck: Symmetry Adaption Reduced to Tabulated Quantities.
    Theoret.Chim.Acta 44,279 (1977)
11. G.Fieck: Symmetry Adaption II.Representations on Symmetric
    Polyhedra. Theoret.Chim.Acta 49,187 (1978)
12. G.Fieck: Symmetry Adaption IV.The Force Constant Matrix of Symmetric
    Molecules. Theoret.Chim.Acta 49,211 (1978)
13. D.B.Chesnut: Finite Groups and Quantum Theory.Wiley,New York,1974
14. M.Rotenberg, R.Bivens, N.Metropolis, K.J.K.Wooten: The 3-j and
    6-j symbols. MIT Press, Cambridge (Mass.), 1960
15. A.R.Edmonds: Angular Momentum in Quantum Mechanics. Princeton
    University Press, Princeton (N.J.), 1960
16. G.Fieck and J.Wirsich: One- and two-particle fractional parentage
    for arbitrary point groups, configurations, and coupling schemes.
    Int.J.Quantum Chem. 18,735 (1980)
17. G.Fieck and J.Wirsich: Electrostatic interaction for arbitrary
    point groups, configurations, and coupling schemes.
    Int.J.Quantum Chem. 18,753 (1980)
18. P.H.Butler: Coupling Coefficients and Tensor Operators for Chains
    of Groups. Phil.Trans.Roy.Soc.(London) A 277,545 (1975)
19. P.H.Butler: Properties and Application of Point Group Coupling
    Coefficients. in: J.C.Donini (Ed.):Recent Advances in Group
    Theory and Their Application to Spectroscopy. Plenum Press,
    New York, 1979
20. S.B.Piepho: Advanced group theoretical techniques and their appli-
    cation to magnetic circular dichroism. in the proceedings cited
    under number 19.
21. G.F.Koster, J.O.Dimmock, R.G.Wheeler and H.Statz: Properties of
    the Thirty-Two Point Groups. MIT Press, Cambridge (Mass.), 1963
22. cf.ref.20, page 411 and ref.19, page 155 (quasi-orthogonal and
    quasi-symplectic).
23. M.R.Kibler and P.A.M.Guichon: Clebsch-Gordan Coefficients for
    Chains of Groups of Interest in Quantum Chemistry.
    Int.J.Quantum Chem. 10,87 (1976)

24. Tang Au-chin, Sun Chia-chung, Kiang Yuan-sun, Deng Zung-hau,Liu Jo-chuang, Chang Chian-er, Yan Go-sen, Goo Zien, Tai Shu-shan: Studies of the Lignad Field Theory. Scientia Sinica 15,610 (1966)
25. E.U. Condon and G.H.Shortley: The Theory of Atomic Spectra. Cambridge University Press, 1967
26. G.Racah: Theory of Complex Spectra III. Phys. Rev. 63,367 (1943)
27. B.R.Judd: Operator Techniques in Atomic Spectroscopy. McGraw-Hill, New York, 1963
28. E.R.Davidson: Use of double cosets in constructing integrals over symmetry orbitals. J.Chem.Phys. 62,400 (1975)
29. L.Jansen, M.Boon: Theory of Finite Groups. Applications in Physics. North-Holland, Amsterdam, 1967 (page 95)
30. RMcWeeny: Symmetry. Pergamon, Oxford, 1963 (page 115)
31. F.A.Cotton: Chemical Applications of Group Theory, 2nd ed. Wiley-Interscience, New York, 1971
32. G.Fieck: Symmetry Adaption III, On the Diagonalization of the Overlap Matrix. Theoret.Chim.Acta 49,199 (1978)
33. cf. for instance: I.Shavitt: Meth.Comp.Phys. 2,1 (1963) or: F.E.Harris: Rev.Mod.Phys. 35,558 (1963)
34. A.P.Yutsis, I.B.Levinson, V.V.Vanagas: The theory of angular momentum. Jerusalem, Israel Program for Scientific Translation, 1962
35. C.C. Roothaan: A Study of Two-Center Integrals Useful in Calculations on Molecular Structure I. J.Chem.Phys. 19,1445 (1951)
36. R.S.Mulliken: Quelques Aspects de le Theorie des Orbitales Moleculaires. J.Chim.Phys. 49,497/675 (1949)
37. H.-W.Streitwolf: Group theory in solid state physics, eq.(15.5) Macdonald, London, 1971
38. F.v.d.Lage, H.A.Bethe: Phys.Rev. 71,612 (1949)
    D.Bell:Rev.Mod.Phys. 26,311 (1954)
39. V.Kopsky: On independent sets of basis functions for irreducible representations of finite groups. J.Math.Phys. 19,574 (1978)
40. cf.appendix of: J.J.Hopfield: J.Phys.Chem.Solids 15,97 (1960)
41. M.Lax: Symmetry Principles in Solid State and Molecular Physics, theorem 3.8.1. Wiley, New York, 1974
42. W.I.Smirnow: Lehrgang der höheren Mathematik, vol. III/1, page 45-49. VEB Deutscher Verlag der Wissenschaften, Berlin, 1964
43. J.Patera, R.T.Sharp, P.Winternitz: Polynomial irreducible tensors for point groups. J.Math.Phys. 19,2362 (1978)
    J.Patera and R.T.Sharp: Generating function techniques pertinent to spectroscopy and crystal physics. in the proceedings cited under number 19.
44. E.Filter: Analytische Methoden zur Auswertung von Mehrzentren-Matrixelementen..., page 32, theorem 2. Dissertation thesis, Regensburg, 1978
45. G.Fieck: Racah algebra and Talmi transformation in the theory of multi-centre integrals of Gaussian orbitals. J.Phys.B (Atom.Mol.Phys.) 12,1063 (1979)
46. G.Fieck:The Multi-Centre Integrals of Derivative Spherical GTOs. Theoret.Chim.Acta 54,323(1980)
47. D.Maretis: Talmi Transformation and the Multicenter Integrals of Harmonic Oscillator Functions. J.Chem.Phys. 71,917 (1979)
48. E.Filter and E.O.Steinborn: Extremely compact formulas for molecular two-centre one-electron integrals and Coulomb integrals over Slater-type atomic orbitals. Phys.Rev. A 18,1 (1978)
49. E.O.Steinborn and E.Filter: Symmetrie and analytische Struktur der Additionstheoreme räumlicher Funktionen und der Mehrzentren-Molekülintegrale über beliebige Atomfunktionen. Theoret.Chim.Acta 52,189 (1979)
50. G.Fieck and D.Maretis: The Tensorial Invariants of Multi-Centre Integrals Generated by the Gradient Method. Unpublished (1978)

51. G.Fieck: Symmetry Adaption III. On the Diagonalization of the Overlap Matrix. Theoret.Chim.Acta 49,199 (1978)
52. W.Kutzelnigg: Einführung in die Theoretische Chemie, vol.2, chapter 5. Verlag Chemie, Weinheim, 1978
53. R.D.Brown, B.H.James, and M.F.O'Dwyer: Molecular Orbital Calculation on Transition Element Compounds I. (cf. footnote 2). Theoret.Chim.Acta 17,264 (1970)
54. H.P.Figeys, P.Geerlings, and C.van Alsenoy: Rotational Invariance of INDO Theories Including d-Orbitals into the Basis Set. Int.J.Quantum Chem. 11,705 (1977)
55. D.N.Nanda and P.T.Narasimhan: On Invariance Requirements in Approximate SCF MO Theory. Int.J.Quantum Chem. 12,215 (1977)
56. B.D.Bird, E.A.Cooke, P.Day, and A.F.Orchard: Derivation and Testing of a Molecular Orbital Description of Ligand Field Spectra. Phil.Trans.Roy.Soc.(London) A 276,277 (1974)
57. P.-O.Löwdin: On the Nonorthogonality Problem. Adv.Quantum Chem. 5,185 (1970)
58. E.P.Wigner: Ueber die elastischen Eigenschwingungen symmetrischer Systeme. Nachr.Ges.Wiss.Göttingen, Math.Phys.Klasse Vol.133 (1930)
59. D.H.Rouvray: The Topological Matrix in Quantum Chemistry. in:A.T.Balaban(Ed.): Chemical Applications of Graph Theory. Academic Press, London, 1976
60. C.J.Ballhausen and H.B.Gray: Molecular Orbital Theory.Benjamin, New York, 1964 and ref.8, pages 393 and 508.
61. C.C.J.Roothaan: New Developments in Molecular Orbital Theory. Rev.Mod.Phys. 23,69 (1951), C.C.J.Roothaan: Self-Consistent Field Theory for Open Shells of Electronic Systems. Rev.Mod. Phys.32,179 (1960), S.Huzinaga: Coupling Operator Method in SCF Theory. J.Chem.Phys. 51,3971 (1969)
62. B.R.Judd: Second Quantization in Atomic Spectroscopy. Johns Hopkins, Baltimore, 1967
63. B.G.Wybourne: Classical Groups for Physicists, section 2.2.10-13. Wiley, New York, 1974
64. E.R.Davidson: Reduced Density Matrices in Quantum Chemistry. Academic Press, New York, 1976
65. W.Kutzelnigg: Ueber die Symmetrie-Eigenschaften der reduzierten Dichtermatrizen... Z.Naturforsch. A 18,1058 (1963)
66. J.Gerratt: Valence Bond Theory. in: Theoretical Chemistry, Vol. 1 (1974), page 60
67. W.Mofitt: Atoms in molecules and crystals. Proc.Roy.Soc. A 210,245 (1951)
68. S.L.Altmann: Induced Representations in Crystals and Molecules. Academic Press, London, 1977
69. R.Dirl: Induced projective representations. J.Math.Phys. 18,2065 (1977)
70. R.Dirl: J.Math.Phys. 20,659/664/671/679 (1979)
71. D.B.Litvin and J.Zak: Clebsch-Gordan Coefficients for Space Groups. J.Math.Phys. 9,212 (1968)
72. L.Trlifaj: Simple Formula for the General Oscillator Brackets. Phys.Rev. C 5,1534 (1972)
73. J.Dobes: A new expression for harmonic oscillator brackets. J.Phys. A 10,2053 (1977)
74. A.W.Niukkanen: Transformation properties of two-particle states. Chem.Phys.Letters 69,174 (1980)
75. P.H.Butler: Point Group Symmetry Applications. Plenum Press, New York, 1981

List of standing abbreviations

AO	atomic orbital
BRM	bicentric reduced matrix element
CFP	coefficient of fractional parentage
CI	configuration interaction
dim...	dimension of ...
GEO	geometrical factor
GTO	Gauss type orbital
Is	isoscalar (factor)
LCAO	linear combination of atomic orbitals
MO	molecular orbital
NSR	non-simply reducible
ord...	order of ...
PIs	polyhedral isoscalar (factor)
PRM	polyhedral reduced matrix element
QRM	quadrocentric reduced matrix element
QSALC	quadrocentric SALC
RME	reduced matrix element
s.-a.	symmetry-adapted
SALC	symmetry-adapted linear combination
SR	simply reducible
STO	Slater type orbital
TRM	tricentric reduced matrix element
TSALC	tricentric SALC
VB	valence bond
WET	Wigner-Eckart theorem
Z(...)	number of ...

Index

The numbers refer to pages, the numbers in brackets to special equations.

Real-Space Renormalization

Editors: **T. W. Burkhardt, J. M. J. van Leeuwen**
1982. 60 figures. XIII, 214 pages
(Topics in Current Physics, Volume 30)
ISBN 3-540-11459-9

Contents:
T. W. Burkhardt, J. M. J. van Leeuwen: Progress
and Problems in Real-Space Renormalization. –
T. W. Burkhardt: Bond-Moving and Variational
Methods in Real-Space Renormalization. –
R. H. Swendsen: Monte Carlo Renormalization. –
G. F. Mazenko, O. T. Valls: The Real Space Dynamic
Renormalization Group. – *P. Pfeuty, R. Jullien,
K. A. Penson:* Renormalization for Quantum
Systems. – *M. Schick:* Application of the Real-
Space Renormalization to Adsorbed Systems. –
H. E. Stanley, P. J. Reynolds, S. Redner, F. Family:
Position-Space Renormalization Group for Models
of Linear Polymers, Branched Polymers, and
Gels. – Subject Index.

G. Eilenberger

Solitons

Mathematical Methods for Physicists
1981. 31 figures. VIII, 192 pages
(Springer Series in Solid-State Sciences, Volume 19)
ISBN 3-540-10223-X

Contents:
Introduction. – The Korteweg-de Vries Equation
(KdV-Equation). – The Inverse Scattering Transfor-
mation (IST) as Illustrated with the KdV. – Inverse
Scattering Theory for Other Evolution Equations. –
The Classical Sine-Gordon Equation (SGE). –
Statistical Mechanics of the Sine-Gordon System. –
Difference Equations: The Toda Lattice. –
Appendix: Mathematical Details. – References. –
Subject Index.

Solitons

Editors: **R. K. Bullough, P. J. Caudrey**
1980. 20 figures. XVIII, 389 pages
(Topics in Current Physics, Volume 17)
ISBN 3-540-09962-X

Contents:
R. K. Bullough, P. J. Caudrey: The Soliton and Its
History. – *G. L. Lamb Jr., D. W. McLaughlin:*
Aspects of Soliton Physics. – *R. K. Bullough,
P. J. Caudrey, H. M. Gibbs:* The Double Sine-
Gordon Equations: A Physically Applicable System
of Equations. – *M. Toda:* On a Nonlinear Lattice
(The Toda Lattice). – *R. Hirota:* Direct Methods in
Soliton Theory. – *A. C. Newell:* The Inverse Scat-
tering Transform. – *V. E. Zakharov:* The Inverse
Scattering Method. – *M. Wadati:* Generalized
Matrix Form of the Inverse Scattering Method. –
F. Calogero, A. Degasperis: Nonlinear Evolution
Equations Solvable by the Inverse Spectral Trans-
form Associated with the Matrix Schrödinger
Equation. – *S. P. Novikov:* A Method of Solving
the Periodic Problem for the KdV Equation and Its
Generalizations. – *L. D. Faddeev:* A Hamiltonian
Interpretation of the Inverse Scattering Method. –
A. H. Luther: Quantum Solitons in Statistical
Physics. Further Remarks on John Scott Russel and
on the Early History of His Solitary Wave. – Note
Added in Proof. – Additional References with
Titles. – Subject Index.

M. Toda

Theory of Nonlinear Lattices

1981. 38 figures. X, 205 pages
(Springer Series in Solid-State Sciences, Volume 20)
ISBN 3-540-10224-8

Contents:
Introduction. – The Lattice with Exponential Inter-
action. – The Spectrum and Construction of Solu-
tions. – Periodic Systems. – Application of the
Hamilton-Jacobi Theory. – Appendices A–J. –
Simplified Answers to Main Problems. – Refe-
rences. – Bibliography. – Subject Index. – List of
Authors Cited in Text.

 Springer-Verlag Berlin Heidelberg New York

Lecture Notes in Physics